The Procurement and Management of Small Works and Minor Maintenance

The CHARTERED
INSTITUTE OF
BUILDING

The Procurement and Management of Small Works and Minor Maintenance

The Principal Considerations for Client Organisations

Jeremy Headley & Alan Griffith

LONGMAN

Addison Wesley Longman Limited
Edinburgh Gate, Harlow
Essex CM20 2JE, England
and Associated Companies throughout the world

Co-published with The Chartered Institute of Building through
Englemere Services Limited
Englemere, Kings Ride, Ascot
Berkshire SL5 8BJ, England

First published 1997

British Library Cataloguing in Publication Data
A catalogue entry for this title is available from the British Library

ISBN 0-582-28873-8

Set by 30 in 10/12 Plantin and Helvetica
Printed and bound in Great Britain by Bookcraft (Bath) Ltd

Contents

List of figures

Preface

The Procurement and Management of Small Works and Minor Maintenance: The Principal Considerations for Client Organisations addresses the key issues surrounding an area of construction and property management that has been frequently overlooked in the past but which is of great importance to both the construction industry and to those utilising its services – the *small works* sector of construction output. The book examines the subject in the context of the large client organisation. Many client organisations occupy large and often diverse property estates which require significant expenditure to facilitate maintenance, alterations, refurbishment or small-scale new-build works. Effective organisation and efficient allocation of resources is absolutely essential to ensure that the works are carried out with maximum success.

The book is arranged in nine chapters that provide a detailed introduction to the subject of small works procurement and management within the large client organisation. Chapter 1 presents an introduction which reviews the key issues in small works procurement and management. Chapter 2 focuses on the definition of small works, identifies their key characteristics and the risks involved in their undertaking. Chapter 3 places small works and minor maintenance in the context of the large organisation and reviews the composition, value and significance of workload linked to clearly determined client policies and objectives.

Factors in framing a procurement and management approach are examined in Chapter 4 where the influence of the building estate function and the relationship of the building assets to the organisation's core activities are significant. Chapter 5 focuses on the considerations surrounding in-house or external procurement, reviewing the engagement of consultants and pre-qualification of contractors. Procurement approach is the subject of Chapter 6. The chapter looks at procurement options, common contractual arrangements, time, cost, quality and geographic considerations and the effect of fluctuations in the economy.

Chapter 7 concentrates on identifying the most appropriate types of contractor for undertaking small works, highlighting the significance of the marketplace and the contractors which operate within it. Management approach is the subject of Chapter 8 and focuses on the control of the key variables and the management thrust for different types of works. Chapter 9 brings the book to a conclusion by highlighting the principal considerations for the successful procurement and management of small works and minor maintenance.

The pace of change in modern large organisations is ever increasing. Management, in general, must be responsive to the more volatile and often more hostile environment by seeking improved organisation and management in all aspects of their business. Small works and minor maintenance is an important aspect of business for the large client and therefore only by implementing a small works procurement and management system that is based on well-defined and measurable objectives will success be maximised. This book is intended to take the reader through the main stages in the formulation and implementation of such a management approach.

Jeremy Headley & Alan Griffith

Acknowledgements

The authors wish to acknowledge the contributions of information by construction industry professionals, and thank all those involved in the production of this book.

Chapter 1

Introduction

Small works and the focus of this book

The intention of this book is to address the principal issues surrounding an area of construction and property management that has frequently been overlooked in the past but which is of great importance to both the construction industry and to those utilising its services – the small works sector of construction output.

Small works, as a grouping, is important both because of its great contribution to total construction output and also because it comprises the largest number of individual items of work. Current estimates place the contribution as lying between 11 and 19 per cent of total construction output.[1] This equates to an annual expenditure of approximately £10 billion in absolute terms. The report *Building Britain 2001*[2] suggested that the economic significance of the repair and maintenance sector of the industry, the sector containing most of the small works activity, is set to increase in proportion to the output of the other sectors in the period 1989–2001. A rise of over 24 per cent is forecast for this period while growth in the new-build sector is estimated to rise by less than 15 per cent. Furthermore, these figures were compiled before the downturn in new construction activity prevalent during the time this book was written. Consequently, the trend for the increasing economic significance of small works is likely to be reinforced by both clients and contractors, attaching increased importance to this sector in the years to come.

Chiefly, because of the relatively small size of the individual items of work comprising this sector of construction work, it has received little attention in the field of published literature. It is both timely and purposeful to redress this situation from the separate perspectives of clients, consultants and contractors as the recessionary conditions prevalent at the time of writing have led to a reduced demand for large construction projects from clients, forcing contractors to seek new opportunities for work. Small works is one of the few areas for which predictions of increasing output can be made with confidence.

From the client's perspective, small works should be treated with greater regard given the financial pressures to remain in existing buildings prevalent in the early 1990s, a situation very different to that of the late 1980s when the demand for new-build progressed as high as business confidence allowed. Continued occupation of an ageing building stock, beyond the point when moving to new premises would be the preferred option in healthier financial times, requires that an increasing amount of maintenance work be carried out to the building fabric to at least maintain tolerable functional and occupational standards. It is probable that internal alteration work will also increase to enable existing buildings to cope with changing requirements in the use of the space they contain, in line with management change, advances in technology and the ever more volatile and hostile financial environment in which the majority of organisations must operate today.

Little is known of the problems of managing small works in situations where there is a high volume of this class of building work, especially in the context of the large property estate, although it is true that many of these problems can be imagined. If such guesswork fields a perception where clients' technical staff are hard pressed to keep on top of the workload and the finance department is inundated with invoices to be paid at the end of every month, that would often be an accurate perception of small works procurement and management. The workload of a local authority maintenance department, for example, may involve around 200 contracts and in excess of 60,000 works orders a year and this serves to illustrate the magnitude of the problem in many cases. There is no doubt that certain ways of procuring small works in a high-volume situation will be inherently more efficient and effective than others. The question is which and how best to approach the design of the small works procurement and management approach?

The starting point for the better understanding of small works is the assumption that more efficient and effective small works management will result from a greater understanding of the attributes of this class of building work. Since small works management has been comparatively ignored in the past by client and management organisations, there is no widely accepted set of guidelines on what constitutes good practice. An early objective in the course of writing this book was to study the range of procurement options currently employed by large client organisations so that the current best practice might be identified. The extent to which effective management practices are employed could then be determined.

As little, if any, research work has been conducted in the small works subject area, certain fundamental aspects need to be addressed initially to provide the basis for a complete understanding of small works. The need to formally define and categorise the particular types of building work comprising small works is paramount as this sets the scene for understanding the concepts and principles of small works procurement and management. The influences of the individual client organisation must be determined with regard to such considerations as size and nature of property estate, geographic dispersion or concentration and closeness of the building stock to the core business. The central importance of clearly defined goals and objectives must be appreciated along with aspects of in-house and external procurement and management. Contractor selection for small works must be understood along with the supervision and control of the workload, which are essential to the likely success of small works procurement and management.

From this book, the reader should gain a purposeful insight into concepts, principles and practices of small works procurement and management. The following specific aspects are prominent:

- To present a formal definition and categorisation of small works.
- To highlight some of the particular problems of the organisation and management of the particular client organisation's characteristics on the range of procurement options.
- To emphasise the importance of taking into account the objectives of the client organisation in formulating a small works procurement policy.
- To identify the decision factors involved in choosing between in-house and external procurement and management of the small works workload.
- To identify the range of procurement options open to client organisations and discuss the design of the procurement approach.
- To present outline guidance for good control and effective supervision of the workload.
- To give basic guidance on the essential requirements for effective management of small works in the context of the large organisation.
- To promote greater management awareness of the issues involved in small works.

This book focuses upon procedures and practice relating to small works of a building nature and with a focus upon the building fabric. Small works in connection with building services or plant facilities represent too specialised an area for inclusion within the scope of this book. Such items of work are characterised by a greater degree of planned maintenance and often more sophisticated procedures generally. However, this does not necessarily imply that in the future there should not be an attempt to bridge the gap between building fabric maintenance and plant and services maintenance. On the contrary, there is much in common between the two areas of maintenance and a case could be made to bring some of the accepted practices of plant and building services maintenance into line with building fabric maintenance.

It is well appreciated that small works and minor maintenance are often intertwined with and caused by the failure of and alterations to building services systems. The decision to exclude this area of maintenance from this book owes more to the fact that most organisations currently draw a distinction between plant maintenance and building fabric maintenance and treat the two areas quite differently in terms of procedures and practice, even though there can be considerable overlap and many small works result as a consequence of services problems.

Small works, for the purposes of this book, does not include refurbishment work. Refurbishment projects tend to be more costly, more resource intensive and generally involve many separate types of construction work carried out by multiple trades. They therefore require a greater degree of planning and control. Refurbishment projects occur much less frequently than items of small works and tend to have a greater degree of consultant input. Subcontractors are common in refurbishment work but not in most small works situations, especially at the lower end of the scale. The increased degree of complexity and the greater number of parties involved in refurbishment work usually demands procurement along more traditional lines with a standard comprehensive form of contract and orthodox tendering procedures, although it is fair to say that this may often be the case for certain types of small works also.

It is also not an intention of this book to review small works in the context of major new-build projects. However, it is recognised that a means of reducing the requirements for building maintenance and small works is to consider these in the design and commissioning of new buildings. Building designs do not always consider the needs of the user with respect to maintainability and, therefore, savings made in the capital cost of new buildings will likely be lost later in the additional expenditure for maintenance, repairs and alterations during the operational life-cycle. The concept of constructability and how small works may be influenced at the design stage is a specialised area that lies outside the scope of this book and is, in fact, covered in another work.[3] In general, however, a client should be forward thinking to anticipate and plan for the building's likely maintenance requirements and future potential use and operation.

Key issues in procurement and management

The key issues in the procurement and management of small works in the context of the large organisation stem largely from the traditional methods of organising and managing the workload. Traditionally, the process would be for senior management of the small works/minor maintenance department to deliver the required working environment and degree of support to assist the building occupiers in the achievement of their organisational goals. This requires an understanding of what their customers require in terms of service provision within the organisation and an awareness of the work required to be done in order to maintain the property portfolio in line with the organisation's objectives.

Having carried out this exercise, the organisation's property management department must then establish the resources implications of providing the requisite level of service provision; that is, the required financial and personnel resources. A proportion of the personnel resources will be in-house specialists, managerial and technical staff, in addition to which there might be an element of directly employed labour. Any resources not available in-house will have to be procured externally from the marketplace, which gives rise to the need for systems and procedures designed to obtain these external resources and to ensure that optimal use is made of them, thereby facilitating the attainment of organisational objectives.

The approach outlined above might seem, prima facie, to be quite straightforward but in reality many factors conspire to place considerable challenges on the smooth running and chances of achieving optimum success in the process. In small works management the characteristics of a very high number of individual jobs of a relatively low value can lead to many management difficulties; the main problem often being that the amount of management time spent administering the small works workload may be disproportionate to the costs of the work itself. In the large organisational context, contractor approved lists can easily run into hundreds and works orders and invoices into thousands annually. This presents a considerable management challenge and indeed a situation in which inefficiences can easily arise.

The situation might be perpetuated by the nature of small works themselves, which often serves to disguise their true importance to the organisation. Consequently, many of the problems inherent in small works may never be fully addressed in practice. Perceptions of small works in the large organisational context require to be changed in order to effect an improvement in management practices. In particular it should be realised that although a job of small works may be relatively unimportant to the organisation's success, the aggregate effect on the organisation of many hundreds if not thousands of such jobs is a significant contributor to the large organisation's continued success, particularly in view of the high cost of small works when the individual jobs are added together.

The traditional small works procurement and management process allows little opportunity to address the issues highlighted since the planned maintenance, alteration and improvement work and small new works element of the workload will require a high management resource input. The little time that might otherwise be available will be consumed by the inevitable reactive maintenance work which arises incessantly and which, along with planned small works, will usually occupy management time fully; the greater the maintenance component of the total workload, the less management time will be available to address the important issues in small works, on both the strategic and operational levels. In such an organisational climate, improvements in procedures will not be realised and the status quo is likely to be merely maintained.

The pace of change in modern large organisations is ever increasing. Management in general must be responsive to the more volatile and often more hostile environment by constantly seeking better methods of construction. This precept should apply equally for small works managers. They should be able to foster continued improvement and seek increased value for money by optimising the balance between the efficiency of resources use and the effectiveness of the end result in a way that is responsive to the ever-changing needs of the organisation. Only by implementing a small works procurement and management system that is based on well-defined and, preferably, measurable objectives will there be any certainty that this ideal is being achieved. This book aims to take the reader through the main stages in the formulation and implementation of such a management approach.

References

1. Griffith A 1992 *Small building works management* Macmillan, Basingstoke
2. University of Reading 1988 *Building Britain 2001*
3. Griffith A and Sidwell AC 1995 *Constructability in building and engineering projects* Macmillan, Basingstoke

Chapter 2

Definitions and characteristics

Small works and minor maintenance defined

Presenting a precise definition of small works can be problematic. This is due to the divergence of opinion about its nature and composition, which arises from the different perspectives of those connected with the small works sector of the construction industry. Definitions of small works, depending on the original source, may vary considerably in criteria such as scale and value, the latter being the most common source of variation.

That small works do exist as a distinct class of building work is indisputable. This chapter proposes a basis for defining small works in a more meaningful way than in terms of value alone, thus overcoming some of the differences in definitions stemming from distinct perspectives. As such it constitutes the foundation for the perception and understanding of small works procurement and management (see Fig 2.1). Small works will comprise:

- Any item of work, either improving, maintaining or altering a part of an existing building.
- Any small-scale new-build operation.
- Any item of maintenance work from any of the categories of maintenance work specified in BS 3811:[1]
  - Planned maintenance.
  - Preventative maintenance.
  - Scheduled maintenance.
  - Condition-based maintenance.
  - Corrective maintenance.
  - Emergency maintenance.
  - Unplanned maintenance.

Since many small-scale items of building work are of a maintenance nature, it is helpful to present alternative categorisations for the range of maintenance-type operations. The above list of types of maintenance can be shortened by restricting it to those most commonly recognised by practitioners on an everyday basis:

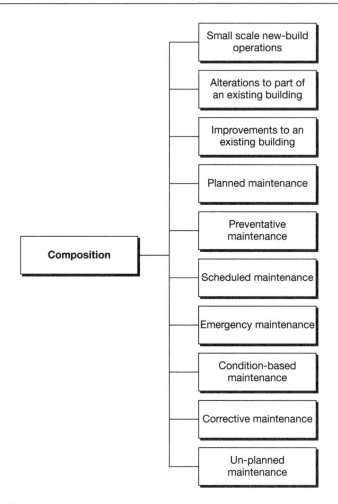

Figure 2.1 The composition of small works

- Routine (or cyclical) maintenance.
- Preventative and corrective maintenance.
- Emergency maintenance.

It may be expedient to reduce the list of types of maintenance work even further, this being possible by specifying only two generic types – *strategic* and *tactical*. Strategic repairs would include both planned and reactive maintenance operations, for which forecasts of future needs can be made and priorities identified. Strategic repair and maintenance work encompasses major structural repair work, such as re-roofing. Tactical maintenance and repair work, on the other hand, is restricted to the day-to-day repair work required immediately, the classic example being the broken window. However, it is important to appreciate that some tactical work might be of strategic importance – a roof collapse being a case in point. For most organisations the majority of tactical maintenance work will be at the lower end of the small works spectrum while much of the strategic maintenance work will be at the higher end.

Key characteristics of small works

Small-scale maintenance work, alterations, improvement work and small new works have many properties that differentiate them from larger building works (see Fig 2.2). These include:

- Many of the individual items of small works within the total workload will be of a maintenance or alteration/improvement nature. These are generally subject to more change once underway than is usual for new work and demand a rapid response when variations occur, e.g. where asbestos is encountered unexpectedly.
- The percentage costs of administration and management in small works, expressed as a proportion of the cost of the work itself, can be much greater than they would be for higher value building work. This is particularly significant at the lower end of the small works spectrum.

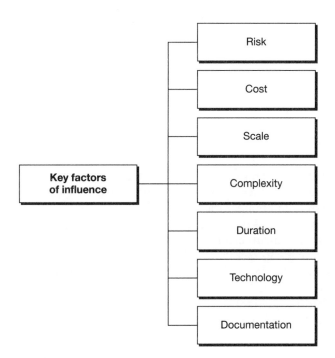

Figure 2.2 The key factors of influence in small works

- The complexity of a typical item of small building work may be much less than that of higher value building work, resulting in a reduced requirement for consultant input and less documentation.
- The contractor will not normally be able to incur high supervisory and managerial costs for low-value work – which makes small works a less attractive proposition for medium-sized or large contractors who inevitably incur much higher head office and supervisory overheads than do the smaller building firms (although larger firms may often gain an advantage over small firms by way of higher discounts from suppliers) and will not be awarded the job in a competitive market unless the client is prepared to pay a premium.
- The duration of an item of small works is commonly much shorter than higher value work.
- Much small works activity takes place in occupied buildings where the considerations of causing minimal disturbance and health and safety regulations are of paramount importance.
- Small works usually involve small quantities of materials, with reduced discounts from suppliers, and a low number of labour tasks.
- Owing to the high level of uncertainty attendant at many small works operations, it is often difficult to specify the extent of the work prior to commencement on site.
- Subcontracted work is less common in small works than building work of a greater magnitude owing to the relatively few different labour tasks involved.

Categories of small works

Three distinct subgroups can be identified under the small works umbrella. These are jobbing works, ordered works and minor building projects (see Fig 2.3).

Definitions for these subgroups have been proposed by Griffith: [2]

- **Jobbing works** – Works carried out to instruction, but without written quotation, without documentation and without a formal agreement between the client/employer and builder/tradesperson.
- **Ordered works** – Works that are too large to adopt a jobbing works approach yet are insufficiently large to justify the use of a standard shorter form of building contract, but nevertheless require a structured approach to procurement and management and utilise documentation and a formal written agreement between client and contractor.
- **Minor building projects** – Those building works procured under a standard shorter form of building contract.

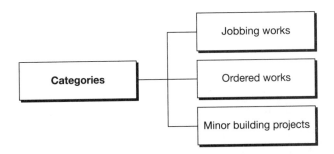

Figure 2.3 The three categories of small works

The relationship between the subgroups of small works can be understood in the form of a small works spectrum, incorporating a continuum of increasing risk, value, complexity and other factors (see Fig 2.4). As a general rule, as the value of the work increases, the degree of formality between the client and contractor will rise commensurately. The exception that proves the rule is the job of significant magnitude but of low complexity, with no perceived requirement for fully comprehensive contract documentation. This situation might arise in various circumstances; for instance, where a contractor, having successfully completed a particular item of work, is retained by a client to carry out further works and the climate of trust is such that the full tendering and contractual process is not undertaken. In developing definitions there is a need to leave sufficient flexibility in their formulation to allow for such special cases, which are not uncommon in small works in the context of the large client organisation.

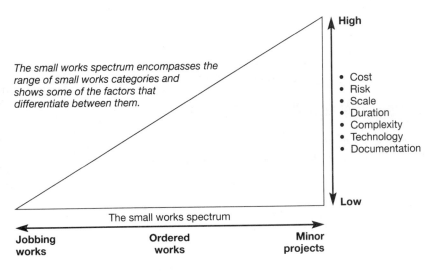

Figure 2.4 The small works spectrum

Dwelling on definitions of small works and its component subgroups might seem like a question of semantics, but a comprehensive treatment of this aspect is a necessary first step not only to define the scope of this book but also to enable a complete understanding of small works management in general. What is clear is that any hard and fast definition of the composition of small works will always be a base of contention. The guiding principle must be that clients should always understand the precise nature of each and every item if the small works are to be managed successfully.

Risk and uncertainty

Risk constitutes a means of discriminating between small works jobs in terms that are becoming more and more the norm for construction work in general – where the analysis of risk is an essential tool in the repertoire of all those concerned with construction management. Risk can be defined as:

The exposure to the chance of injury or loss – in construction terms this is generally financial, physical or time-based exposure to injury or loss.

Uncertainty may be defined as:

The extent to which an outcome is not known or not known certainly, not to be depended on or is changeable.

Construction work will always involve an inherent degree of risk and uncertainty. This is just as true for small works as it is for larger work. It should not automatically be assumed that the lesser degree of complexity and magnitude associated with small works means reduced risk to the client. On the contrary, the failure of an item of small works may have profound consequences on a critical function or process within an organisation. The consequences of any risk should always be measured in terms of the indirect costs of the work as well as the direct cost. Risk should always be managed.

Risk management in small works can be defined as:

The reduction of the balance of risks and uncertainties associated with the estimation of various outcomes in small works management to an acceptable level.

Four components (see Fig 2.5) may be identified as being the main sources of risk, within which specific sources of risk for small works associated with each component may be identified:

● **Threats** – These comprise the major influences which might lead to an unfavourable outcome. Threats arise from contracting out the work, cost overruns, late delivery, poor quality of finished product and scant regard to health and safety legislation.

- **Resources** – All those factors which might be adversely affected by the threats facing them. Resources include building occupiers, the organisation's revenue income, the property assets of the organisation, time of the persons charged with administering the small works workload.
- **Modifying factors** – The attributes which are able to influence the chance of the threats being realised or the magnitude of the consequences. Selection of procurement route and contractual arrangements are crucial influences of small works risk management. The procurement and contract strategies and the management of their implementation are major modifying factors.
- **Consequences** – The effects of the realised threats on the resources. These may include management time wasted 'fire-fighting' poor quality work with money squandered on remedial works, excessive disturbance to building occupiers, processes or functions, contractor insolvency, contractor default or work not being carried out with the required degree of urgency.

The process of risk analysis constitutes a discipline which seeks to identify worst case scenarios and formulate plans on that basis. It depends on asking 'what if?' questions, quantifying the consequences of the possible scenarios and, finally, drawing up contingency plans to either eliminate or lessen the effects of the realisation of the scenarios. The steps in risk analysis for small works can be summarised as follows:

- Identify the major risks inherent in each and every small works job by reference to a checklist and by knowledge of the particular project or building.
- Supplement this process by discussion with stakeholders in the work item – that is, all those persons, responsible for processes or functions, who might be affected by either the job itself or its non-performance.
- Refer to databases which keep track of the frequency of occurrence and consequences of risks that have been realised in the past.

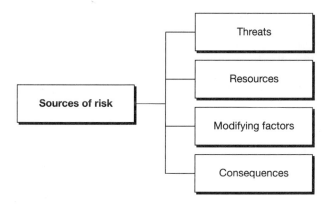

Figure 2.5 The main sources of risk in small works

- Where the job warrants it, notional best case and worst case prices may be arrived at by reference to information held on databases – these prices should take account of both direct and indirect costs of the job (in practice this may be deemed to be worth while only for the more significant jobs).
- Plan courses of action and formulate contingency plans on the basis of the above process.

The risk analysis procedure outlined might be argued to be no more than a formalised version of a procedure that is usually adopted in the normal course of small works management. This may be so in many cases but, in terms of achieving greater effectiveness, there is no comparison between a rigorous and methodical system that achieves consistency in application and a more *ad hoc*, perhaps intuitive, procedure.

Review and discussion

Small works may be categorised according to various conventions. Most organisations do in fact recognise that different categories of small works exist. However, the categories often owe more to somewhat arbitrary or inherited traditional groupings than to classification according to the distinct characteristics of the item of work. For instance, classifying works according to value bands alone, where works in particular value bands will have different procurement routes, is a common situation.

A common policy might be along the following lines: works under an estimated value of £1,000 may be procured without competition by approaching a contractor from the select list to carry out work up to this value; works of an estimated value of £1,000 – £10,000 must be procured competitively with a minimum of three contractors from the select list; and works above an estimated value of £10,000 must be procured competitively under a JCT Minor Works, or similar, contract. Works categorised in this way may be procured by inappropriate means which may not take account of the uniqueness of every item of building work. Also, it is rare for the value bands to keep pace with inflation so there is a tendency for the bands to lose weight over time, and it is not uncommon for the value bands to remain unchanged for as long as a decade.

It is possible to demonstrate the inappropriateness of a small works procurement policy along the lines described, where the selection of procurement route is on the basis of estimated cost of the work alone, by drawing a parallel with the current usage of the JCT Minor Works contract. The increased and successful usage of the JCT Minor Works contract, for certain types of project with values much higher than that originally proposed as the practical upper limit for its use, clearly illustrates the validity of looking at all aspects of a particular item of work prior to selecting the procurement route. This leads to the need for a sufficient number of alternative procurement routes to provide the required degree of flexibility to cope with the uniqueness of every individual item of construction work.

Yet another flawed aspect to a procurement policy based on value bands alone is the fact that the guidelines are often ignored in practice. Let us suppose, for example, that in the case of the procurement approach described, the individuals responsible for procuring an item of work, of an estimated value of say £1,200, decided, for whatever reason, that they do not want to tender the work competitively. It is fairly common practice in situations such as this for the work to be awarded as two separate jobs of, say, £800 and £400, on a non-competitive basis to the selected contractor. It should be stressed that this practice is usually resorted to for good reason and is not restricted to the private sector alone. The point is, that there is little to be gained from having a rigid policy if it is ignored or misconstrued in practice.

In the absence of regular reviews of the organisation's management of the process, it is far better to have a less rigid policy which has a sufficient degree of flexibility built in to cope with the particular requirements of each eventuality. Such a degree of flexibility could best be provided by a policy which enabled the selection of the procurement route to be made on the basis of all the characteristics of an item of work, not on one criteria alone.

It is the contention of this book that a simple definition of small works according to either type of work or cost of the job alone is insufficient. What is required is a more holistic means of looking at any individual item of small building work due to the unique nature of individual jobs. In other words, there are many interacting influences arising from the specific characteristics of any item of small works which require to be recognised. This view demands an approach which looks at all of the separate influences imposed by the circumstances of the job prior to final selection of the procurement route. Without such an approach it is likely that the smooth running of the small works management approach and ultimately the management of small works will be jeopardised and the frequency of occurrence of disputes will likely increase.

References

1. BS 3811 1984 *Maintenance management terms in terotechnology*
2. Griffith A 1992 *Small building works management* Macmillan, Basingstoke

Chapter 3

The large organisation

Types of client

Chapter 2 presented a case for a fresh perspective on small works and minor maintenance. This originates from the fact that there are several distinct categories and many types of small works, these being derived from the scale of each job, ranging from jobbing works to minor projects, and the type of work involved – maintenance, new-build, alteration or improvement work. Each category comprises particular features yet, paradoxically, there exist a sufficient number of common factors between the subgroups identified for the general distinction between small works and larger works to hold true. The important point is that small works is an area with many unique facets of organisation and management which must be recognised by the management policy adopted by the particular client. It is not sufficient to frame a policy around one criteria, such as value bands, alone. The small works management policy must have sufficient in-built flexibility to cope with every eventuality arising from the unique nature of each item of construction work.

Chapter 3 moves on from this position to put small works in the context of the large client organisation where particularly challenging demands are placed on the staff responsible for small works procurement and management. The larger the organisation the higher is the small works workload and the more critical this area becomes if optimal resource utilisation and maximum effectiveness are to be achieved. Expectations in large organisations are generally higher as well. The need for the management approach to support the organisation's core business needs is also of fundamental importance.

There is a great diversity in types of client. Apart from the obvious distinction between private and public sector clients, they can be classified according to their experience of, and expertise in, construction. A further level of classification is possible according to the extent to which the client derives income from the property stock directly. A primary constructor would gain most income from the construction and leasing of property while a secondary constructor would require buildings merely to support the activities of the organisation's core business.

Whereas many small clients have only a limited understanding of construction, the majority of large client organisations have a reasonable degree of expertise in building procurement. Consequently, the large client's expectations in terms of finished product performance and delivery are usually much higher than those of smaller clients. Large clients, who have a consistently high construction work demand, will commonly have a large amount of expertise retained in house, although recent trends are moving away from this. This in-house expertise will generally comprise construction professionals at senior management level, whose job, in part, is to formulate and implement policy. They will usually be supported by personnel in a more junior capacity, some of whom will procure and manage the ongoing small works workload.

Diversity between client organisations also arises through the uniqueness of individual property estates. The estate of the typical large organisation is a complex entity comprising a disparate range of building types. The buildings are often a broad mixture, including purpose-built from new, adapted, converted, refurbished, leased or owned. They may be distributed across a wide area, within a particular region or local area, or may be concentrated on one site. They will invariably have been built during different time periods using various methods.

Although large client organisations are unique in many ways, they have attributes in common. The challenges on the personnel charged with small works management arise from the high frequency of individual items of work required to maintain every large property estate, irrespective of its particular characteristics – whether geographically compact or diverse, whether owned or leased, whether predominantly city centre or out of town, whether mostly new, old or a mixture of several types of construction.

Why is the wide scope for variation between different clients significant? Simply because it is extremely unlikely that there will be one all-embracing formula suitable for implementation by each and every organisation with regard to the management of the small works workload. Individual clients' systems and procedures will require some degree of tailoring to suit the particular characteristics of the organisation, since each will have different criteria for success.

One further aspect of client organisations is worthy of comment. The internal environment in which clients' in-house property departments must operate is not always ideal. The climate of many large client organisations has been suggested to be one of policy making by committees. This aspect is considered by Milne.[1] In such climates effective control is often in the hands of the finance branch. It is suggested that in these conditions there often exists a tendency to maintain the status quo, whether this is a conscious decision or an unfortunate consequence of the particular circumstances. The status quo more than likely comprises a set of traditional procedures and practices that have evolved haphazardly over a long period of time in tandem with the increasing size, or changing composition, of the organisation's building estate. Given this situation, it is hardly surprising that minimal attention has been paid to seeking ways to improve the efficiency and effectiveness of small works procurement in many organisations.

Organisation and management

Facets of small works (see Fig 3.1) that highlight the need for increased attention on management practices in the area include:

- **The disguised importance** – A common misconception held by many construction professionals is that small works is not a terribly onerous area, when compared with larger projects for instance – it is also less glamorous. However, the burden of a high frequency of smaller scale jobs may amount to more in financial terms than the cost of any project, and may be more difficult to administer, in terms of the absence of the managerial rigour and disciplines associated with the management of projects.
- **Total workload** – The volume of work in the category of small works in the typical organisation is very much greater than that of larger or higher value work (for example, the local authority maintenance department with a work throughput of around 200 contracts and 60,000 works orders a year illustrates the significance of this point).
- **Perceptions** – Importantly, small works is often an area which plays second fiddle to major projects within many organisations. Although this is an understandable situation, is it right? Although the cost of the average small works job may be low, when the total amount expended on this class of work is calculated, it often exceeds the expenditure on major works in many organisations. It may come as a surprise to learn that the average value of maintenance work in many organisations often amounts to only a few hundred pounds, whereas a figure well into the thousands may have been expected.
- **Cost of failures** – Mistakes are expensive, embarrassing and often avoidable. *Ad hoc* management or loose control systems can contribute to failures. An ineffective and expensive small works management approach is a failure.

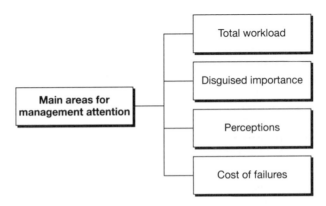

Figure 3.1 The main areas for management attention in small works

Notwithstanding the need to address the above, it is often found that many large organisations fail to recognise certain fundamental characteristics in the management of small works. Without an appreciation of these it is unlikely that either the efficiency of resource use or effectiveness of the management effort will be optimised.

In the context of the large estate, the nature of jobbing works is that they are generally repetitive – that is, basically similar items of work will be carried out many times in various locations throughout any given time period. There is a trend, therefore, towards a routine or operations management situation at this end of the scale. At the opposite end of the small works spectrum, however, the tendency for individual jobs will be to move towards greater uniqueness. There will be relatively fewer similar jobs and the criteria for success will be more clearly defined. The trend here is towards project management (see Fig 3.2). These two opposing influences are central to an understanding of the organisation and management of small works. The design of the small works management approach should take account of these characteristics in order to be successful.

It is worth contrasting works at the lower end of the small works spectrum, i.e. those of a routine or operations nature, with those at the upper end, i.e. those tending towards projects. Routine small works operations will be characterised by their repetitive nature and the relatively stable environment in which they take place, compared to the project environment. The climate for routine operations is subject to gradual evolution of change, whereas projects are more noted for radical and sudden change. Projects rely on temporary teams of goal-oriented people, whereas operations are dependent on stable groups with well-defined roles and goals.

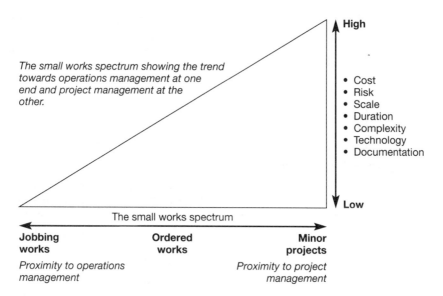

Figure 3.2 The orientation of management within the small works spectrum

The definition of a project proposed by Turner[2] draws the distinction between projects and operations:

A project is an endeavour in which human material and financial resources are organised in a novel way, to deliver a unique scope of work, of given specification, within constraints of cost and time.

Turner suggests that if 'in a novel way' were deleted and 'unique' replaced with 'repetitive', this might be a suitable definition for routine operations.

The small works workload

The large estate will require many hundreds if not many thousands of individual jobs to be carried out every year. Most of the items of work carried out will be of a small scale. This smaller scale of individual items of building work often serves to disguise the true significance of small works in the context of large property estate maintenance. Although the value of a particular item of small building work may appear to be fairly trivial when compared with that of larger building works, the total amount expended annually on this class of works by a client with a large building estate may be very significant indeed – it is often seen that it is not exceptional for the annual expenditure on small works to be of a similar magnitude to that spent on larger construction work. This results in pressure to seek ever greater efficiency in the area.

For most organisations property is second only to staff payroll in terms of cost, and by a considerable margin. There is a gradual shift in attitude away from the traditional view of the property estate as a liability in recognition that the organisation's property assets will often represent 30 per cent, or more, of its total asset value. The majority of the costs associated with building ownership and occupancy are fixed; however, a sizeable proportion are controllable. There is a growing realisation among building owners and occupiers that the controllable costs represent a profit opportunity. A more effective and efficient small works procurement policy represents one area of the controllable costs which can assist in the drive to optimal use of the property estate.

Composition of the workload and value

Before an operation can be controlled, it must be capable of being measured, otherwise how is one to know whether the control mechanisms are successful or not? In the context of small works, it is often particularly difficult to quantify the extent of the problem. The numbers of works orders may be easily enough ascertained, but what about the total expenditure within value bands or on particular types of work? What is the exact quantification of the direct and indirect costs associated with an item of work, or indeed across the total workload?

However, managers of small works must be ever mindful that, in measurement, certain classes of works may be inappropriately lumped together.

As detailed a knowledge as possible of the composition of the total small works volume is required for an understanding of how it might be managed more effectively. This requires that:

- Firstly, the necessary management information systems are in place – the required degree of detail and the size of the workload in large organisations requires the use of the latest computer systems to enable this level of information retrieval to be attained.
- The volume of work be broken down into such categories as are appropriate to the individual organisation, but always being mindful of the distinction between the main subgroups of small works.
- The management time spent on administering each category of work be recorded.
- A means of ascertaining good and poor performance is in place.
- Both the direct and indirect costs associated with the various categories of small works under different forms of procurement are understood.

Importance of policies and objectives

In framing the small works management approach, the starting point is a clear expression of the *raison d'être* of the small works department originating from the overall premises policy. To this end, the small works department should formulate a mission statement from which basis the department's goals may be derived. The proximity of jobbing works to operations management and minor projects to project management has an influence on the setting of goals. A project will have project-specific goals whereas routine operations will have operations-specific goals. It is worth clarifying the difference between goals and objectives at this point. Goals are general statements of intent while objectives constitute quantifiable, and therefore measurable, expressions of the goals.

Once well-defined objectives have been identified, strategies for their realisation can be formulated. This is followed by actions to implement the strategies. Control is exerted by monitoring the actions assessing the success of the actions in achieving the objectives and by making such modifications to strategies and actions as are necessary in order to enhance the achievement of the objectives.

Clients' criteria for success, manifest in the goals and objectives set by them for particular types of work, are of two main kinds:

- Primary goals
- Secondary goals

Most commonly, the former comprise the ranking of time, cost and quality in the appropriate order for a specific job or customer. Secondary goals, however, are also an important consideration for most clients. These are more numerous than the primary goals and some of the possible goals are listed below under their 'parent' primary goal:

- **Time**
  - Rapid response and commencement on site.
  - Certainty of finish date.
  - 24-hour and 365-day cover for certain work.
- **Cost**
  - Price competitive.
  - Price certainty.
  - Attainment of value for money.
  - Minimise cost of mistakes.
  - Minimise staff resource costs.
  - Minimise direct cost of work.
  - Minimise indirect costs of work.
- **Quality**
  - Different standards for front of shop/behind the scenes.
  - Appropriate for purpose.
  - Guarantee for client.
- **Other**
  - Health and Safety at Work considerations.
  - Flexibility.
  - Ease of data gathering.
  - Security.
  - Accountability for decision-makers.

The presence of multiple goals, for both routine operations and projects, means that a certain amount of conflict will often be present between them. Effective management requires a means of prioritising the goals to overcome this potential for conflict.

Review and discussion

There are often two opposite forces at work in most client organisations. The pressure to reduce the headcount, by concentrating on the core business and outsourcing certain functions, which must be balanced against the need to maintain sufficient influence over decision-making to maintain effective control over the operations.

Most organisations exist in increasingly hostile and turbulent environments: technological, political, economic, etc. The only way that they can survive and prosper in this climate is to be more adaptable and flexible in the face of change.

This need impinges on all aspects of the organisation's operations, including its attitudes to its property estate. In the context of the management and administration of the small works workload, this same need for flexibility and adaptability in the face of change is carried through. This means that any written policy related to the small works procurement policy should have an intrinsic element of flexibility so that it will maintain its currency over time and in the climate of an ever-increasing rate of change.

Many organisations have either recently carried out, or are currently conducting, an examination of the effectiveness of their current small works management policies. This suggests that there has been widespread internal recognition of inefficiencies in the ways that many large organisations procure their small works as well as changes in the construction marketplace. This trend has resulted in many organisations carrying out re-engineering and new policy implementation exercises in an attempt to overcome the perceived problems.

The catalyst for the trend to reappraisal of the effectiveness of current approaches includes such common problems as:

- The large amount of paperwork generated by some traditional systems of maintenance procurement have made these systems both unwieldy to operate and tedious for those persons charged with their administration.
- The pressure to seek greater efficiency in the organisation generally due to the current economic climate. The effects of this range from a desire to reduce the headcount in the maintenance department to a desire to be more in control of the expenditure.
- The longer term effects of repeatedly postponing maintenance expenditure in the past are beginning to make themselves more apparent by greater fluctuations in the maintenance workload. This has resulted in such problems as a spate of major flat roof repairs or replacements as a long-term consequence of the temporary patch repairs carried out previously. This tends to overstretch the often limited resources of many maintenance departments.
- The contractor approved lists administered by many large organisations have been allowed to become far too extensive in many cases. Lists compiled on an *ad hoc* basis have a tendency to grow considerably over time – contractor lists of 50–70 or more individual contractors for one location, or much larger lists for regional areas, are time consuming and difficult to both set up and operate properly. This has resulted in a desire to minimise the points of contact to as few contracting organisations as possible in order to ease this administrative burden in many cases.
- Where a large organisation has permitted its regional managers a degree of autonomy in recent years in the ways that they implement the organisation's maintenance procurement policy, this has resulted in several distinct policies existing within the same organisation. A desire to bring all regions into line has been encountered where this trend has become apparent.

New strategies which have been recently implemented or are currently under investigation include:

- Placing the responsibility for the management of the organisation's maintenance workload entirely in the hands of external management contracting or facilities management organisations.
- Wider use of term contracts with the aim of reducing the administrative workload.
- The appointment of trade or specialist contractors to carry out the total workload of that trade or specialism on a national basis, e.g. national glazing contractors. Because few contractors engaged on this national basis have the resources to provide complete coverage, they may subcontract local contractors – often those previously held in the client's contractor list – to do much of the work. By this means the required degree of cover is achieved and the client retains many of the advantages of having work done by contractors with whom they have had wide experience in the past.

Common elements of the new policies implemented include:

- A desire to ease the unnecessary element of the administrative burden of managers.
- Restructuring of a maintenance department which is more responsive to organisational change – this is an increasingly important factor with the frequent changes wrought by external and internal influences in recent times in most large organisations.
- Greater unification of policy across the entire organisation, where there might have been a degree of regional variation previously.

Small works departments often play second fiddle to major projects. But is this appropriate? Does not the role of the small works department play a greater part in enhancing both the value of a particular property asset and the satisfaction of the personnel working within it? If this is the case then the question of relative status in the organisation must, or at least should, be reappraised.

Some organisations have added other responsibilities to that of the maintenance department, such as seeking out and developing new business opportunities. This seems to be an effective way of adding interest to the job function, retaining staff and increasing status in the organisation. It is also useful in bringing the activities of the department to the attention of executive management.

References

1. Milne RD 1985 *Building estate maintenance administration* E & FN Spon, London
2. Turner R *What are projects and project management?* Henley Working Paper HWP 2/90, Henley Management College

Chapter 4

The procurement and management approach

Influence of the building estate function

In the case of building works of sufficient size and complexity to require the use of a standard comprehensive form of building contract and full tendering procedures, it is generally recognised that clients can gain many benefits from the selection of the most appropriate method of procurement for a particular project – 'procurement' meaning the framework under which the design and construction of the project is purchased and controlled.

As well as the different financial considerations and various degrees of risk attendant to each available procurement option, certain routes will be more certain to achieve specific goals and objectives than others. The selection procedure must reflect these differences in emphasis. If this axiom is true for higher value building projects, then the same considerations should apply in the selection of the most appropriate procurement method for smaller building works. This may be reflected in the possible savings that might accrue from the employment of more efficient and effective procurement routes, although it is stressed that *most appropriate* is not synonymous with *cheapest*.

There are many alternative procurement routes for small building works and the selection procedure should ensure that the most appropriate route is chosen – the best option being that which is most likely to achieve organisational objectives. This chapter highlights some of the more important factors underlying the design of the client's small works procurement system.

The typical property estate comprises buildings with a broad mixture of functions – from the corporate head office, or administrative centre, to those buildings provided to house the core activities of the business or businesses of the organisation directly. Larger organisations will inevitably occupy many different buildings which often accommodate many different functions.

Criteria for success for individual buildings and functional uses of those buildings may comprise various:

- Quality standards.
- Service standards.
- Frequency and nature of work arising.

The building function is important because each type of building represented in the client's portfolio will have particular requirements for its successful management, including the management of its small works. An example of this may occur where, for example, the property estate of a large and diversified organisation comprises department stores and retail food stores. In this context a retail food store can, essentially, be a chain-store operation, and the way in which the buildings are developed, managed and maintained reflects this.

The nature of chain-stores is such that there is usually an ethos and discipline imposed from the centre or parent organisation on the way this part of the business is run so that there is an almost identical performance by each store, within broad terms. With this approach, such aspects as the way in which the buildings are operated, serviced and maintained is clearly determined. The visual impression created in such retail food stores is, more often that not, consistent from one store to the next. Moreover, the life-span of a retail food store is likely to be considerably less than other types of store as a result of frequent change in customer preference and the high levels of competitiveness in the retail food sector.

In contrast, a department store is quite different. Department stores are much more complicated buildings which stand alone to a great extent within a mixed property portfolio. Whereas a retail food store might be around 30,000 square feet gross, a department store might be 250,000 square feet gross. The visual impression projected in a department store is geared towards the local community, the seasonal retail cycle and other influences. Essentially the department store is a stage-set which is usually less susceptable to change and therefore has a longer life-span than a retail food store.

Notwithstanding, a property portfolio will likely have buildings of all ages – ranging from some stores with medieval parts over 100 years old to new ones – and therefore, the technical and managerial problems are varied. A building might have cast-iron columns and timber beams while another might have a substantial amount of prestressed concrete in its structure. All buildings will have discrete objectives regarding the approach to small works which reflect the specific nature of the business, the building's age, its type and its composition.

Building assets and core activities

For certain types of large organisation the property stock is central to the core business; for hotel groups, for example, it could be argued that the buildings themselves are the core business. Other types of large organisation rely on their facilities only to the extent that they enable the major activities to be carried out; in this case there may be a greater tendency for maintenance budgets to be trimmed, sometimes to the extent that only essential work is carried out. It would be extremely rare for large hotel groups to allow their facilities to decay to the same extent as has been evident in the estates of some local authorities, for instance.

This aspect has an impact on the management of small works and will affect such decision factors as:

- Provision of direct labour.
- Choice of procurement route.
- Proportion of planned maintenance work in total maintenance workload.
- Quality standards.
- Service standards.

The main factor for most of these decisions will be the extent to which the income-sensitive aspects of the organisation's activities are affected.

Impact of quality assurance

Quality has become the watchword for more and more organisations, both strategically and operationally, with the advent of Total Quality Management (TQM). BS 5750,[1] in its various parts, promulgates the international and European standards – ISO 9000 and EN 29000[2] respectively – for quality systems in connection with the appointment of suppliers and purchasers. It is applicable to all organisations that design, manufacture, construct or install products or who provide services, and has served to raise the general awareness of the duties and responsibilities of persons able to influence the achievement of high standards of quality in the provision of goods and services.

Many large organisations have identified quality assurance as a means of enhancing both the perception of the organisation by its customers and the quality with which it carries out its core business activities.

Quality assurance is defined in BS 4778: Part 1 in the following terms:

All those planned and systematic actions necessary to provide adequate confidence that a product or service will satisfy given requirements for quality.[3]

Its main intention is to eradicate substandard performance and poor quality. From its beginnings in the engineering and manufacturing industries, as a means of improving the quality of products by reducing the incidence of defects, its effects have permeated through to organisations in many sectors of construction and throughout functions at every level within those organisations.

Quality systems must be adopted wholesale throughout an entire organisation to be effective, so where they have been implemented, all parts of the client organisation have been affected by some degree of change associated with the increased attention focused on quality. Clients' small works and minor maintenance sections, departments or divisions are no exception to this.

In the context of building work in general over the last 10 to 15 years, quality assurance has been adopted relatively slowly by contracting companies. Traditional approaches to quality management have, until recently, been predominantly *ad hoc* and although in-house quality assurance schemes are common among most large and some medium-sized contractors, full third-party certification under BS 5750 among contractors in general is today more widespread but not total.

In terms of small works procurement by large client organisations, it can be assumed that a quality assured client will prefer to engage the services of a quality assured contractor, other things being equal. Apart from anything else, this should relieve the client from much of the effort involved in the often onerous process of contractor pre-qualification. Contracting organisations of all sizes will see the continuing uptake of the tenets of quality assurance by clients as a motor for change in their own firms to achieve a marketing advantage over their competitor companies.

Full third-party certification under BS 5750/ISO 9000 essentially demonstrates that the certificated company has implemented a recognised and authoritative quality system – 'the organisation structure, responsibilities, activities, resources and events that together provide organised procedures and methods of implementation' (BS 4778) – which has been assessed by an accredited third party to comply with the relevant part of BS 5750/ISO 9000 and associated Quality Assessment Schedule. Lesser levels of assessment may also be carried out for organisations not adopting full certification – internally by the organisation itself or externally by a purchaser (first-party assessment and second-party assessment, respectively).

The certification procedure is based on a periodic assessment of the functioning quality system and, as such, enhanced quality of the product or service is not guaranteed by the implementation of the standard. It does, however suggest a system which certifies that a firm meets the required standards. A criticism often levelled at BS 5750/ISO 9000 is that the standard is not wholly applicable to the construction industry. While this may have some validity, such points should not detract from the advantages of an implementation of BS 5750 /ISO 9000 since the standard does provide a framework which may be adapted to suit particular circumstances and within which the risks of occurrence of both defects and poor service provision should be minimised. For detailed guidance on the implications of quality assurance within construction, the reader is directed to the references presented in the select bibliography at the end of the book.

Statutory legislation

Clients must always be aware of the impact of statutory regulations on their activities since these will undoubtedly influence the criteria for success of their small works management approach. This is especially so for large clients to which a greater number of regulations generally apply. With Europe becoming ever more integrated in pursuit of the single market, European directives and standards will have an increasing impact on the operations of large organisations.

Health and safety is a particularly important aspect of small works management, especially because much work of this nature will be carried out in occupied buildings. Whether occupiers are members of the public or the client's own staff, it is imperative that sufficient health and safety safeguards are incorporated in the client's small works procurement and management strategy. New legislation will invariably mean higher costs to clients in ensuring compliance with the full requirements of the regulations already in force or coming into force. This will require a reappraisal of existing procedures in the light of the increased responsibility placed on clients to ensure compliance with the legislation, and also in such areas as contractor pre-qualification and record keeping.

Clients should brief all new contractors and operatives on their health and safety aspirations and policy prior to starting work. The costs of this process alone might well be significant if repeated continually, which might well mitigate against certain small works procurement options.

Guidelines for the implementation of health and safety legislation in small works procurement and management, in general, include:

- The main legal duties and responsibilities of employers, managers and employees should be understood.
- Objectives should be set that are both achievable and measurable.
- Plans should be developed and implemented to achieve these objectives.
- Health and safety hazards should be identified and measures for their control formulated.
- A risk assessment exercise should be undertaken to establish the risk of occurrence of the hazards identified.
- Work activities should be organised to minimise the risks from human, jobs and organisational factors.
- Employees should be provided with sufficient training, information, instruction and supervision to carry out their work with minimal risk.
- All accidents should be investigated to determine the immediate and underlying causes so that appropriate remedial and preventative action can be taken.
- All communications between higher management, safety professionals, enforcing officers and staff at all levels regarding health and safety matters should be carried out in an effective manner that will lead to appropriate action.

Importance of procurement form

Generally, for most large organisations, buildings do not exist for their own sake. They can be seen merely as factors of production that make some contribution to the achievement of the goals of the organisation owning, occupying or renting them. If they make no contribution, the question must be asked. Why are they there in the first place?

Each building will have specific criteria for its success by which its overall performance in use may be ascertained. These criteria will be derived from such factors as the building usage, ownership, construction, current state of repair, location and capacity to generate income for the organisation. The criteria for success, or objectives, provide management with the means to understand, plan, forecast and control activities effectively. This applies equally to the management of the small works workload, which plays such an essential part in the operation of the building stock.

One of the major influences on the successful outcome of an item of small works, defined as the extent to which the key objectives are realised, is undoubtedly the procurement route. The way that the item of work is purchased from the market, or carried out in house if applicable, can either facilitate or hinder the achievement of the objectives set, depending on the specific circumstances. Clearly specified objectives, relating to such areas as individual building characteristics, nature of the core business, quality assurance and the management of health and safety policy, can constitute a basis for realistic performance assessment to take place.

Review and discussion

Investigations[4] have found that experienced secondary clients – that is, those clients who require buildings to support their core business rather than to derive income from the property stock directly – select their procurement routes primarily on the basis (in order of significance) of:

- Advice from in-house experts.
- Previous experience of similar projects.
- Company policy/financial regulations.
- Advice from external consultants.

It has been said that the heavy reliance on past experience and constraints of in-house policy engender a conservative approach to selection of procurement route. During the course of writing this book, this tendency was seen to be equally valid for small works procurement in many cases.

What is required, in a climate of increasing change and uncertainty that is subject to an increased regulatory burden, is a more flexible and adaptable small works procurement and management approach based on the particular characteristics of the client organisation and clear objectives developing from these.

References

1. British Standards Institution 1987 BS 5750: *Specification for quality systems*
2. International Standards Organisation 1994 ISO 9000: *Specification for quality systems*
3. British Standards Institution 1971 BS 4778
4. Masterman JWE 1992 *Introduction to building procurement systems*, E & FN Spon, London

Chapter 5

In-house or external procurement and management

In-house and external procurement

Not so long ago this chapter might have been concerned merely with the relative advantages and disadvantages of directly employed labour versus external contractors for small works procurement. This approach would pay little heed to the current trend of many large organisations towards 'outsourcing' the management of many functions previously carried out in-house, including the management of the small works workload. The scope of the chapter has therefore been widened to incorporate reference to the expanding field of facilities management.

There will always be good arguments for, and equally valid ones against, the use of in-house directly employed labour. The issues involved are contentious and require careful consideration, especially in the context of the large client organisation for whom direct labour will usually be a more viable option than for smaller clients.

For many organisations the choice between direct labour and contract labour, for certain categories of work, is not clear cut. When the decision has been made to either maintain or establish an element of directly employed labour, the size of the establishment must be calculated with care. Such organisations must ensure that the size of the direct labour force does not exceed that which can comfortably deal with the magnitude of the identified work area existing in periods of reduced construction demand – that is, during troughs in the ongoing workload cycle.

The identified work area, for which the direct labour element is retained, usually encompasses those tasks which are critical to the success of the organisation and whose poor performance would have a negative effect on the generation of income to the organisation, should the required level of service position not be available.

- Direct labour is well suited to carrying out small-scale routine work of the sort that exists at the lower end of the small works spectrum.
- Multi-skilled craftspeople, where appropriate, provide the maximum degree

of flexibility for a direct labour force since rigid job or trade demarcation should not be allowed to develop, although this ideal is often subject to geographical and health and safety constraints.

Current trends

Large organisations have in recent times tended towards the elimination of much of the direct labour that was traditionally retained to carry out the ongoing minor maintenance and repair workload. Some large organisations are able to contract outside labour on a semi-permanent basis for their exclusive use. By this means they, in effect, gain many of the advantages of maintaining an element of direct labour without some of the negative aspects. Clients gain increased flexibility in the face of variable work levels, along with a reduced management workload, and there is a mechanism for tendering the work periodically to gain the most advantageous price available from the market at any time.

Facilities management

It is important to note that there is more to facilities management than merely constituting a suitable pigeon hole to accommodate those areas of managerial responsibility not directly related to an organisation's core business or main activities. The essence of facilities management propounded by the Association of Facilities Managers is that it coordinates the physical workplace with the people and work of the organisation by integrating the principles of business administration, architecture and the behavioural and engineering sciences.

A preferred basis for the definition of facilities management is conveyed in the form of a facilities management mission:[1]

To manage the organisation's accommodation assets, through time, in the most cost effective way to meet agreed business objectives.

This view fits in with the general ethos for this book, which is that a small works management approach should, likewise, be managed through time in the most effective way to achieve agreed objectives. An alternative view of facilities management, apparently held by many senior managers at the present time, sees it is as little more than the management of those activities and functions related to the property and necessary to support the core business or major activities but not directly related to the core business or major activities of the organisation. These activities might include maintenance, cleaning, reprographics and security, among others. The rise of contract facilities management companies which aim to relieve organisations of these burdensome functions provides evidence of the willingness of senior managers to see facilities management in these terms.

Whatever the preferred perspective, the management of the small works and minor maintenance workload should rightly be identified as an important aspect of facilities management. Since the maintenance bill accounts for approximately 8 to 12 per cent of a building's annual running costs, or as much as 30 per cent for an old building,[2] with small works comprising a sizeable proportion of this total, its importance should not be underestimated.

Arguments for and against placing the management of small works in the hands of an external facilities management contractor are given below:

- **For**
  - Specialists should be more in touch with current best practice.
  - Specialists should be more efficient.
  - 'Downsizing' or 'rightsizing' the organisation in response to change should be easier.
  - Performance could be linked to results.
  - Raises the profile of the contribution of property estate performance to the organisation's success in the eyes of organisational decision makers.
- **Against**
  - Potential loss of control.
  - Potentially more difficult to achieve accountability.
  - It can be difficult to achieve a situation of lasting trust between the client and external facilities management contractor, where the facilities manager does not seek to maximise profit but equates his or her success with that of the client and attempts to drive costs down accordingly.
  - The difficulty in defining the requirements of the service provision.

A risk analysis should always be carried out prior to making the final decision as to whether to contract out a function. This should pinpoint all the things that could go wrong by the adoption of this strategy, including loss of response time and deterioration of quality. Ultimately, a balance needs to be struck between the perceived advantages and the things that could go wrong.

Engagement of consultants

There are many reasons why large organisations will choose to engage external consultants, including:

- The lack of specialist or professional expertise in-house.
- Consultants may have particular knowledge of their market and an awareness of 'the state of the art' in their field that the client organisation might lack.
- They provide a means of easing peaks in the workload.
- The overall workload may be reduced.
- Cost savings may be gained over the retention of in-house staff.

For the management of small works, most large organisations will have sufficient expertise available in-house to effectively manage the day-to-day workload. However, there may be situations where specialists are called in for any of the reasons cited above. In this eventuality, consultants should be selected as carefully as contractors. Indeed, consultants often think of themselves as contractors.

A client's selection of consultants should be on the basis of:

- The level of experience of the consultant engaged.
- The seniority of the member of the consultant's staff commissioned.
- Reputation – of both individuals and particular firms.
- The ease with which client's staff can gain access to partners of the firm of consultants.
- Importantly, past experience of working with the consultant.
- Cost, although this might be low on the list of priorities when the commissioning of consultants is concerned (this is manifest in the fact that commissioning does not automatically follow open or selective competition in many cases). However, cost may become the main consideration if problems are encountered with other criteria.

Pre-qualification lists for consultants can be every bit as important as for the selection of contractors.

There is a need to monitor the performance of consultants. This can be quite a tricky exercise given the cost premium paid by clients to certain consultants for specific expertise and, hopefully, a higher quality of service. The most effective means of monitoring consultants' performance is by employing more than one to carry out a certain type of work, perhaps across different geographic areas for those large organisations which are geographically spread.

It should be remembered that by engaging consultants the need to manage has not been eliminated. On the contrary, consultants require to be managed just as do contractors. There should always be a sufficient number of client's staff in place to oversee the activities of consultants and monitor their ongoing performance. Also, apparent cost savings realised by farming out certain aspects of work to consultants may be false, by merely swapping costs from one accounting category to another. Costings of the relative merits of engaging consultants or facilities management firms should always be thorough and should seek to ascertain the true situation. Finally, if the relationship with a facilities management contractor turns sour, it may be difficult to replace that person at short notice. Contingency plans should always be in place to prepare for this eventuality.

Pre-qualification lists for contractors

Small works, by its very nature, is often ideally suited to being carried out by smaller contracting firms, leading to a management problem in the resulting large number of these smaller contractors. Any client organisation must ensure that the comparatively large number of smaller contractors with whom it does business, compared to larger works, are able to carry out the work effectively. To achieve this, sufficient staff of suitable experience must be retained by the organisation to oversee the management process.

The former National Federation of Building Trades Employers (NFBTE)[3] identified the sources of the failure of many contractors to provide the required service to clients in the maintenance and minor works field. The blame for inefficiency was shared between clients and contractors.

The contractors' main faults were identified as:

- Failure to provide a proper written estimate and sufficiently detailed specification.
- Lack of knowledge or disregard of the elementary principles of business administration.
- Failure to keep adequate liaison with the client during the progress of the contract, particularly in regard to possible delays and cost variations.
- Inadequate knowledge of building principles and building regulations requirements.
- The use of inadequately skilled labour.
- Failure to notify the client that subcontract labour is being used, and the reasons for this.
- Inadequate supervision of the work carried out by subcontract labour.
- The use of unsuitable materials.
- Use of unsafe or inadequate and unsuitable plant and tools.

The client organisation must ensure that the process by which it selects contractors minimises the occurrence of such risks as they can have an adverse effect on the success of the management of the small works workload. Properly administered systems for contractor prequalification are a means to this end. If the system is not administered properly its capacity to influence the overall success of the management of the small works workload will be lessened considerably. Although an effective contractor pre-qualification system can involve much management time and expense, these costs should in theory be more than offset by reductions in the occurrence of the problems highlighted above.

The aim of contractor pre-qualification is to compile a register which, once in operation, should ease considerably the problems associated with finding suitable contractors to carry out small works jobs and place less reliance on word of mouth. Each entry in the register should be concise enough to facilitate quick reference, but might appropriately contain as a minimum the following information:

- Reference number.
- Name and address of firm – in alphabetical order.
- Point of contact – telephone numbers.
- Operating area.

- Brief description of firm, including date established, number of employees, plant and equipment owned, facilities owned or leased.
- Judgement on the company's financial standing, updated to include its current position.
- Type of work carried out.
- Areas of particular expertise.
- Summary of work range and number of contracts carried out.
- Performance assessment of most recently completed jobs for the client.

Some of the above categories will not be relevant in certain cases, and as they will also be subject to constant change, regular updating is essential.

The register should not only state the different types of work carried out by contractors, but should also give details of the sizes of their previous jobs or contracts, preferably subdividing work in accordance with the small works spectrum.

The summary information provided on the register will be taken from more detailed data gathered from a variety of sources including questionnaires, contractor enquiries and published sources. This stage constitutes the actual process of pre-qualification. The process can be quite time consuming and if it is repeated continuously throughout a large organisation much duplication of effort may result. A centrally administered pre-qualification system would therefore seem to offer most benefits.

The nature of the work involved in contractor pre-qualification points to the use of computer systems and databases to ease the data assimilation and interpretation process. By this means, amendments can be made easily and quickly as necessary, information can be retrieved speedily and searching can be carried out under many different fields, e.g. location, size of jobs, specialisation.

Review and discussion

The likely consensus between large organisations would seem to be that where an element of direct labour already exists in the organisation this is a good thing and an appropriate, and usually minimum, level should be retained. The converse is true of organisations that do not have direct labour provision and prefer to keep it that way.

Within the retail sector of the economy, some large client organisations have found it advantageous to establish teams of mobile technicians to cover stores within various geographic areas. These technicians tend to be electricians for the simple reason that the consequences of fresh produce chiller cabinets breaking down can be very expensive with the potential for thousands of pounds worth of stock to be ruined. A fast reaction to the inevitable breakdowns in plant that do occur is thus guaranteed and the technicians may be employed on preventative maintenance duties when not called upon for urgent and emergency duties. The theory lying behind this approach may be equally applicable to building fabric work for some large organisations in certain circumstances. Where there is an income-sensitive area of the organisation's activities, the provision of an element of direct labour to service this can often make much sense.

References

1. Thomson T 1990 The essence of facilities management. *Facilities* **8**
2. RICS 1989 *The maintenance of commercial property* GCPPA/Report (89)9
3. NFBTE 1978 Small Builders' Section: *Managing a smaller building firm*, National Federation of Building Trades Employers

Chapter 6

Procurement approach

Procurement options

The large client, or the small works management agency external to the client organisation, having formulated key objectives for each category of small works, must then design a procurement approach that will facilitate the achievement of these objectives (the objectives should be subdivided into two distinct types: operations objectives for the routine small works jobs and project objectives for the more substantial small works jobs – the latter may be project specific). The term 'procurement approach' rather than 'contract approach' is used advisedly because in small works there is often a distinct lack of formality in certain arrangements.

The selection of procurement route in small works management is important because it sets up the framework and degree of formality within which the client and contractor must work together. It is obviously desirable if a harmonious and cooperative working relationship can be established at the start and maintained thereafter. This should reduce the potential for disputes to become acrimonious and damaging.

The procurement route selection process is important because it is one of the major influences on the ultimate success of a job and is also one of the first decisions made affecting the undertaking of the work. A wide range of procurement options is available to the client who wishes to employ contractors to undertake the work, any one of whom might be suitable for a particular item of small works in certain circumstances. In selecting the appropriate procurement route option for an item of work, the most important criterion will normally be the complexity of the job. This should have a greater bearing on the decision-making process than the value of the work, for instance.

- Degree of in-house expertise available.
- Desire to use in-house management or outsource the management of the workforce.
- Size of the organisation.
- Core business.

- Formality and number of levels for decision making.
- Source of finance.
- Distinct features of the client business.
- Nature of the work itself.

In terms of the nature of the work, the characteristics (see Fig 6.1) that will point to the appropriate procurement approach and contractual arrangement are:

- **Type**
  - Complexity.
  - Design input.
  - Management input required.
- **Scale**
  - Cost.
  - Extent to which scope of work can be defined precisely at tender stage.
  - Degree of financial management required.
- **Volume**
  - Amount of similar work in the overall workload.
  - Benefits accruing from economies of scale.

Common arrangements for procurement

The arrangements in common usage among client organisations (see Fig 6.2) include:

- Works order.
- Daywork order.
- Daywork term.
- Measured term.
- Minor measured term.
- Annual term.
- Minor lump sum fixed price.

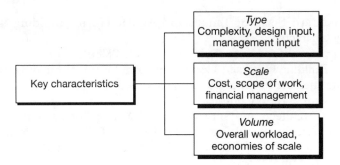

Figure 6.1 The key characteristics of small works

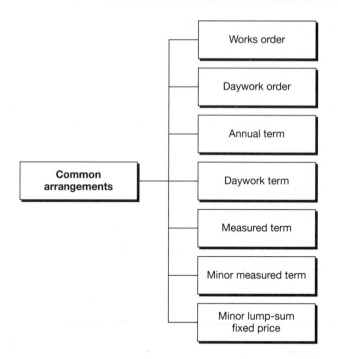

Figure 6.2 The common arrangements for small works procurement

It is not intended to explain the intricacies of each of these options, since they have been dealt with more than adequately in other publications, but merely to point out that a wide range of options is available, some of which have scope for tailoring to suit particular circumstances. Such refinements might include the empowerment of building users to procure certain types of work or 'bulk' contracts by which otherwise unconnected items of work are procured under one arrangement leading to possible economies of scale.

Available forms of contract

- JCT Standard Form of Tender and Agreement for Building Works of a Jobbing Character.
- JCT Agreement for Minor Building Works (MW80).
- Scottish Building Contracts Committee form for minor building works.
- JCT Intermediate Form of Building Contract.
- JCT Measured Term Contract.
- Faculty of Architects and Surveyors (FAS) Minor Works Contract.
- Faculty of Architects and Surveyors (FAS) Small Works Contract.
- General Conditions of Government Contract (GC) for Building and Civil Engineering Minor Works.
- DoE General Conditions of Contract.

Time, cost and quality considerations

Rightly or wrongly, time, cost and quality implications are often overridden by other decision factors as the basis for procurement route selection in small works. Examples of decision factors which might override time, cost and quality considerations include:

- Year-end spend, by which unexpended element of budgets may enable low priority work to be undertaken.
- Easiest procurement route to administer, regardless of time, cost and quality considerations.
- Availability of suitable contractors.
- Term contract in place with pressures to allocate work to term contractor, especially in times of reduced overall workload.
- Direct labour on hand with pressure to allocate certain amount of work to them.

If, however, time, cost and quality are the main criteria for the selection process, then advocating any particular procurement route as being most likely to achieve any particular criterion in any but the most general terms would be fraught with difficulty. This can be seen as the nub of the problem of small works procurement route selection since the particular circumstances surrounding each item of work will dictate the appropriate strategy for each job. For instance, it might be tempting to advocate direct labour as the best means of enhancing work quality, but there should be no reason why an external contractor, probably with a history of doing work for a particular client, should not produce an equally good standard of work, or indeed a superior standard. It is often the achievement of the correct balance of time, cost, quality and other objectives that will point to the best procurement option.

Procurement form and categories of work

It is suggested that a certain degree of tailoring of the small works procurement approach to accommodate the particular characteristics of each category of work carried out is both appropriate and necessary. If this approach seems unduly complicated, this is only in recognition of the complexity of small works itself. Since the characteristics of each client organisation have such an important influence on the requirements for a procurement approach, there can be no best solution and a contingency approach is necessary. Each organisation must tailor its own bespoke solution to its small works procurement approach based on both its own characteristics and the facets of small works generally.

Geographic considerations

For those client organisations that do not have the luxury of being based entirely within a compact and clearly defined region, the management of the small works workload is complicated by the distribution of their centres of operation. Certain organisations, such as some banks and high street retailers, will be represented by branches located throughout the length and breadth of the UK. Organisations such as these must take account of the varying geographic conditions likely to be encountered, from the Shetlands to the Channel Islands, in the formulation of their small works procurement and management strategies. Variable factors include:

- Availability of suitable contractors in the locality.
- Cost of labour, plant and raw materials.
- Regional variation in construction materials and methods.
- Nature and incidence of building defects encountered across different areas.
- Local economies.

As an example of geographic considerations, a large UK company, owning property assets in most large towns and cities in the country, adopted quite distinct strategies for the management of its small works and minor maintenance, taking into account varying demography. It saw considerable benefits from implementing term contracts as part of a policy of reducing its staff resource workload and minimising the staff headcount of the maintenance department. However, the company saw this policy as being most appropriate only for areas of highest population density. Different approaches were adopted in other demographically defined areas with the result that the company effectively implemented four separate strategies for the management of its small works nationally (see Fig 6.3).

A further illustration is the example of a large organisation with 4,000 properties widely distributed around the UK. The company had five regional properties acting as the administrative and technical organisation for each region. Traditionally, each of the 4,000 properties procured its own maintenance contractor to meet its individual small works requirements. The company recently introduced term contracts to its regional organisations on a trial basis, and initial results showed considerable promise in comparison with the traditional approach. It was considered that term contracts would be most appropriate in specific large conurbations where most of the company's properties were situated in localised areas, while individual contractors would continue to be procured to service the small works requirements of the more remote properties.

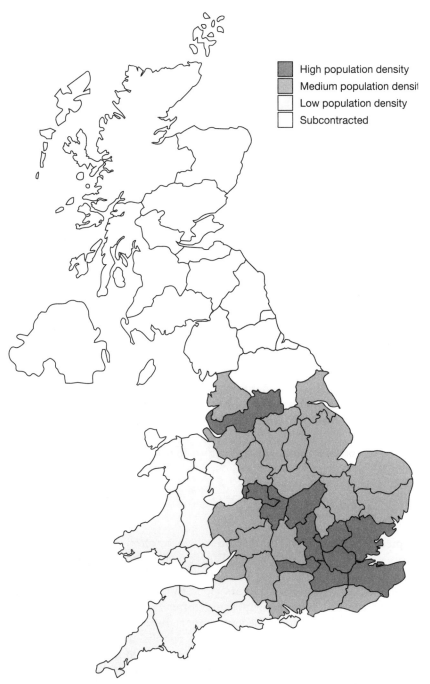

Figure 6.3 The small works procurement strategy of a certain large insurance organisation's property portfolio essentially consists of four distinct substrategies, in recognition of regional and demographic factors

Effect of fluctuations in the economy

Small works has in recent years become an increasingly important component of total construction output in the UK. This is largely due to the downward trend in the proportion of new work and the upward trend of maintenance work as components of the total. In 1972, repair and maintenance work constituted only 27 per cent of total construction output. This proportion had increased to 40 per cent of the total by 1986.[1]

The UK economy has been characterised by periodic booms and slumps for many years. Construction is usually one of the industries most affected by trade cycles but the effects on the repair and maintenance sector tend not to be as severe as those on new work. In the small works context, large clients are affected by slumps in the economy in the following ways:

- There will probably be less money available for small works in general, with maintenance budgets trimmed in particular.
- Contractors' tender prices are keener, so directly employed labour might be a less attractive option.
- There will be a greater incidence of contractors going bankrupt or into liquidation during and immediately after any economic downturn.
- When 'the green shoots of recovery' appear following recessionary conditions, contractors will often increase their capacity to take on more work by subcontracting labour. This can be a source of problems.
- Larger contractors, who might not be interested in carrying out small works during periods of full order books, will compete for the available maintenance and repair work with smaller contractors.

All of these considerations will impact on the procurement approach to some extent.

The consideration of security

The piecemeal nature of small building works often poses difficulties where security considerations are relevant. Where a large number of small contractors must be retained to carry out work when required for the client, there will inevitably be a stream of contractors' operatives and others with a need to gain access to the work location, which may often be in a secure area. Examples where security considerations become important include air transport facilities, telecommunications installations, military establishments and corporate head offices. The short duration and large number of individual jobs across the entire estate mean that this is always going to be a potentially weak point in an organisation's security strategy. It is simply not possible for client's staff to supervise a multitude of small jobs thoroughly enough to ensure watertight security where a multitude of contractors are engaged.

Measures that can be taken to minimise the risk of security breaches to clients include:

- Briefing contractors as to the client's security policy and requirements.
- Insisting that each contractor nominates a responsible person to safeguard the clients' security considerations on site.
- Instituting rigid reporting procedures for contractors' arrival and departure on site.
- Briefing the client's own security staff as to the location of contractors and the progress of the work.
- Electronic access controls.

Direct labour, where each operative can be vetted before being offered a job, offers the best solution to the security aspect. However, in circumstances where directly employed labour cannot be afforded or is not required permanently and where a stable labour establishment is necessary, an alternative would be to contract-in a labour force under a daywork term arrangement. Prominent examples of this are common in the chemical and petroleum industries where contractors are employed under three- or five-year term contracts over many years as staff and operatives are known and trusted by the client organisation.

Accountability for decision making

In the past, there have been unfortunate instances where client's staff with responsibility for procuring maintenance work from contracting firms have been implicated in corrupt activities. Successful prosecutions have resulted in terms of imprisonment. Clients must always be aware of the danger of corruption and should guard against its occurrence by period checks on decision makers at the interface between the client organisation and contractors. Multiple levels of authorisation for contract administration will also serve to create a culture where there is less scope for the unscrupulous.

Review and discussion

The correct choice of procurement route for any job of small works is a major influence on its ultimate success, however that success, may be defined for the particular job. There are two aspects to small works procurement:

- The choice of procurement route for individual jobs.
- The design of the procurement approach.

The procurement route selection process must be carried out in a systematic way to achieve a consistent approach where large clients are concerned and to avoid the introduction of inefficiencies. The design of the procurement approach should ensure that sufficient procurement options are provided to gain the required degree of flexibility, choice and security.

Small works is an area in which the aim should be to strike the right balance between the cultivation of long-term relationships between clients and contractors and the provision of sufficient safeguards to ensure that no more than market prices are being paid. The size of individual jobs often gives rise to attitudes on behalf of clients and contractors where it is not considered worth pursuing claims connected with disputes. The atmosphere could be soured in the long term and therefore all parties, and especially small works contractors, will often prefer to view problems in context of a 'swings and roundabouts' situation.

Reference

1. Briscoe G 1988 *The economics of the construction industry* Batsford/CIOB

Chapter 7

Selecting the contractor

Contracting firms

It is vital to select the most appropriate contracting firm for small works and minor maintenance activities, in terms of size and suitability, for a particular set of circumstances. This is an essential requirement, both to facilitate the smooth running of the procurement system and to minimise the risk to the client of contractor failure.

Contracting firms exist in many different shapes and forms, many of which may be suitable for certain small works situations. They range from the very small to the very large, but whatever the size of an individual firm it will fall into one of the following categories:

- General building.
- Specialist trades.
- Specialist maintenance.
- Building and civil engineering.
- Civil engineering.

It is possible to discriminate between contracting firms in terms of their size as well as by the nature of their business. This criteria is often the major factor influencing the price, risk and business relationship existing in small works procurement. The following three general categories are usually applied in this respect:

- Small firms/jobbing builders.
- Medium-sized firms.
- Large national contracting firms.

In the selection of contracting firms, generalisations on the suitability of generic types of contractor to carry out certain types of work can be hazardous; therefore, the client's procedures for contractor pre-qualification and final selection should always be the final arbiter of a contractor's suitability.

Chapter 5 reviewed the arguments for and against the client employing direct labour or contracted labour. When the client procures any small works or minor maintenance, he or she will have a somewhat predetermined set of objectives on the works requirements. These notional needs are then put into perspective by the consultant – designer, engineer or surveyor – that is employed to transform those ideas into a definite set of building requirements. The problem lies in determining the type and nature of contractor required to carry out the works, if direct labour is not to be used.

The client will need to consider the capabilities of different contractors by differentiating and evaluating their individual attributes. Moreover, the practice of contractor vetting must go hand in hand with the client's small works policy for determining the different types of approach to small works, so both must be considered at the same time. Although the client may have employed a contractor on a previous occasion this should not preclude re-evaluation since small works and minor maintenance are, like other construction activities, always different in some way, and if the client is to obtain the best possible deal the works should always be carefully considered.

The client, in evaluating a potential small works contractor, should consider the following:

- The organisation of the contractor in terms of support structure and administrative capacity.
- The financial base to support the extent of the works over its duration, which is important if the works are to be procured on a term contract basis.
- The contracter's resources to support and sustain the works efficiently and effectively.
- The normal operating limits of the contractor in terms of range, scope and size.

It is important that the client considers these points as, in general, the small works situation is often different from that found in the procurement of larger works. From seeking to answer questions based on the aforementioned points, the client will be able to consider the contractor's reputation and standing within the business community. In addition, one of the most salient points will be determined, that of the contractor's ability to undertake the particular scope of the works – which is important in small works where scale, volume, and specialist skills are often significant.

Differentiating contractors' attributes

The approach to contractor selection should take into account the fact that contractors in the field of maintenance and small works are in a quite different situation from those engaged in larger contracts. It is suggested that in the case of small works, the successful contractor's main emphasis is usually on securing repeat business from larger clients. While many contractors would actively seek repeat business for larger works as well, the higher frequency of individual jobs in small works means that the potential for a close relationship to develop between both parties in small works is somewhat greater. Consequently there is less of a tendency to resort to litigation to resolve disputes.

Contractors often put a higher priority on maintaining a harmonious relationship than on pursuing every possible claim, which might jeopardise this goal. If the contractor feels that he or she has lost out in a dispute over any one job, there will be many more in the future which will mitigate any perceived or actual loss. A good long-term relationship will result from consistently achieving client satisfaction and this will be the successful small works contractor's highest priority. These circumstances often lead to a relationship somewhat akin to a partnership between contractor and client.

Although there are undoubtedly dangers in allowing too cosy a relationship to develop between the client and contractor there will be mutual advantages for both from a more symbiotic association if sufficient safeguards can be built into the arrangement. One result of this symbiosis will be a degree of coordination between client's and contractor's systems in small works, precisely because of the long-term nature of the relationships.

There will be many advantages to this familiarity with each other's systems and requirements. The net effect should be, after a settling-in period, to reduce misunderstandings and minimise the occurrence of disputes. It is the contractor's responsibility to ensure that his or her systems are adapted to suit the individual client. The client can and should aid the contractor in this by giving a clear statment of objectives. A periodic review of procedures to discuss possible procedural refinements will provide a means of identifying and resolving problems as they are seen to develop.

The marketplace

At the present time many contracting firms of all sizes are exploring the small works sector of the market, having identified it as one of the few areas offering opportunities for growth in construction. Clients should be aware of the risk that, in the future, some contractors, having dabbled in small works during the recession, may be tempted to move back to their traditional market of larger projects where they have most experience and where they consider more profit can be made. The influence of the current market cannot therefore be overlooked. Since long-term relationships, and policies and procedures are central to small works procurement, all of the system should be capable of being adapted to the changing trade cycles characteristic of the UK economy in recent years.

The ability to respond to changing economic needs and market demands should always be one of the client's priorities. Proactive clients will be constantly vigilant and 'manage' their building assets to increase the value of their portfolio. An important part of this activity is to maintain and improve their buildings through effective maintenance and small works procurement and management. All these aspects must be capable of being managed within a changing environment and fluctuating marketplace.

Partnering the client

The key consideration in contractor selection is that, ideally, the contractors being considered should be seen to be identifying themselves with the goals and objectives the client sees as most important for the type of work on which they may be engaged. This desirable and often elusive aspiration probably transcends the importance of any policy decisions that might restrict the engagement of contractors for small works by specifying generic types of contractor as only being suitable to carry out work in narrowly defined areas.

It has been identified as highly significant that small works depend, to a great extent, upon the relationship and dialogue that is formed between the client and contractor. This is true of all small works but particularly important where term contract arrangements are used. From a client's perspective it is essential that the chosen contractor is seen as a 'partner' to the client's policy, aims and objectives. It is due to good partnering, perhaps more than any other factor, that clients often employ a contractor on term conditions for many years, where reliability and trust have been firmly established.

Review and discussion

The National Federation of Building Trades Employers guide to *Managing a smaller building firm*[1] listed some major failings of clients in procuring maintenance and minor works from contracted labour. These included:

- Lack of understanding and complete and clear instructions as to the work required.
- Interference with work in progress by giving further instructions, often to the wrong person, and failure to realise that alterations and additions to agreed specification require reallocation of labour, causing disruption to the continuity on their own contract and consequential disturbance to other customers.
- Inability to realise that an unreasonably low estimate is likely to result in unsatisfactory work.
- Changes of mind without checking the likely consequences of these variations.

While it would be almost impossible to completely eradicate the occurrence of these and other factors which lead to inefficiency in the relationship between clients and small works contractors, the potential for disputes arising as a result can be mitigated by careful selection of contractors. Therefore, contractor vetting and the policy for engaging specific types of contractor for the particular works should be considered together.

Contractor selection should be firmly founded in the client getting the best possible deal and obtaining maximum value for money. However, this is not synonymous with the quickest or cheapest route to arrangement but rather, the selection of a contractor who best meets the aims and objectives of the client. It is for this reason that partnering is important in contractor selection. It is essential to the client organisation as trust and reliability is often a prerequisite to the concept of undertaking small works and is essential to the contractor as small works provide a great opportunity for securing repeat business, sometimes over a long period of time. Careful contractor selection is, therefore, to the benefit of both parties.

Reference

1. NFBTE 1978 Small Builders' Section: *Managing a smaller building firm*, National Federation of Building Trades Employers

Chapter 8

Management approach

Supervision of the workload

Supervision is synonymous with control in the client's management of the small works workload – control exerted both on the overall small works management process and aspects of the individual job. Control is necessary because only rarely will plans and actions be entirely successful in their execution without changes. It is in the nature of things, and some might say especially so in construction work, that if something can go wrong it ultimately will go wrong. This adage is as true for small works management as it is for any other management process, whether we are referring to the control of contractors or of any other aspect of the client's procurement and management approach.

It is important to be clear how a control mechanism should fit into the small works management approach. The management process begins with the identification of goals and measurable objectives; at the small-scale end of the small works spectrum these goals and objectives will be operations oriented, while at the large-scale end of the spectrum they will often be project-specific goals and objectives. Plans are formulated and implemented that are designed to realise the goals and objectives.

In small works management in the large organisation the plans will be contained, implicitly or explicitly, in the overall procurement approach. Indeed, the main framework for exerting control in small works procurement is often the procurement approach. The procurement process will produce returns to the organisation, such as the achievement of the required level of service to the core activities of the organisation.

For any small works management system, a control system is required to monitor the extent to which the objectives are being realised and also to trigger appropriate control measures whenever and wherever these are required – which is usually when significant deviations from the plans are detected. The control system is particularly important where the environment in which the small works management takes place is subject to fluctuations.

Control of the key variables

Control must be exerted on two levels in small works management. The first level is control of the overall small works approach and the second is the control of the individual job. Any inefficiencies manifest at the individual job level may often be regarded as being of little importance, especially for jobbing works, but it should be realised that the aggregate effect of procedures which are not wholly effective at the individual job level will be significant if widespread across the organisation. This is why control of the small works management approach is so important if organisational effectiveness is to be optimised. Any control system will have associated costs for its implementation and maintenance, so it is necessary to weigh costs carefully against benefits.

The basic small works control process should rely on the existence of defined levels of performance, specified as objectives, for both individual jobs and repetitive operations; the levels of performance will normally be defined in terms of service standards, quality standards and cost criteria – the main ingredients of value for money. However, along with the *primary* goals of time, cost and quality, it may be desired to achieve certain *secondary* goals.

Secondary goals might include health and safety aspects, minimisation of staff resource costs, etc., the level of attainment of which will have an impact on overall assessments of performance. (The prioritisation of multiple goals is a control process in itself.) Individual jobs important enough to justify project-specific objectives will normally manifest themselves as minor projects. Repetitive operations, the majority of which will be simple and evident as jobbing works, will involve operations-specific objectives.

The level of attainment of the objectives during the undertaking of these activities and processes is measured as the basis for performance assessment. Adjustments are effected as necessary to bring actual performance into line with planned performance.

Small works management commonly experiences two difficulties in respect to control: first, works are often inadequately undertaken at the first attempt, requiring work to be repeated; and, second, some jobs are overspecified and therefore, overworked. In order to guard against the occurrence of these situations, control mechanisms which address the primary goals (time; cost and quality) should be put in place. Various levels of time/cost programming – *long term, medium term, short term* – can be used to achieve the commensurate degree of control to meet organisational and operational needs.

Long-term programming considers the broad vision and overall framework for the purposes of organisational planning, and should:

- Establish the policy and approach to small works management.
- Consider the total small works workload within a set timeframe and cost budget.
- Determine the requirement for in-house and external resources.
- Consider the implications of the scheduled works upon the core activities of the business.

Medium-term programming should:

- Distribute the small works workload continuously and uniformly over the timeframe/budget.
- Consider the work nature, pattern and requirements in relation to the ongoing activities of the business.
- Provide a schedule for design information, tendering procedures and contractor selection.

Short-term programming should:

- Subdivide the medium-term programme into individual small works jobs.
- Allow for time/cost allocation of each job.
- Provide a schedule for small works inspection and progress evaluation.
- Present a mechanism for feedback and review.

It is not proposed in this book to present specific methods for planning and monitoring time and cost in small works management. The development of programmes and financial spreadsheets, together with their various reporting mechanisms, have been the detailed subject of other work. The reader is therefore directed to the reference which follows at the end of this chapter.[1]

Since almost all small works are undertaken in existing and usually occupied premises, consultation with building users is paramount. Secondary goals – for example security and health and safety – must be considered when the works are first planned and when they are carried out to avoid disruption to ongoing business activities. Further situations requiring detailed consultation might be where small works are to take place in environmentally sensitive locations or in buildings of historic or architectural value where statutory consultees or groups with special interests may become involved. In such cases not only will the planning and control mechanisms need to be considered, but also the fundamental policy and approach to the works since both the primary and secondary goals may be significantly different from those of other small works.

In the context of the large organisation, control is often of a preventative nature, whereby the mechanism for control consists of constraints imposed by rules, regulations, policies and procedures. These are often contained in formal training and procedural manuals or they may be informal, in which case they will be communicated verbally. As control in this case does not rely on continuous feedback, it is termed *static* control. Control can also be exerted by the use of feedback loops which enable constant monitoring and adjustment. In this case, the control mechanism is *dynamic*.

Some client organisations will rely on a mixture of static and dynamic control for the management of the small works. Within the ethos of this book, it is considered that the dynamic form of control mechanism should be accorded at least as great an importance as the static mechanism since this is more sensitive to changes in the organisational environment. This represents a change to traditional practice whereby the static form of control has often been considered the most important small works control mechanism.

Control may be effected by means of:

- Static control systems.
- Dynamic feedback mechanisms.
- Some combination of dynamic and static systems.

The control of small works procurement in the large organisational context is commonly based on a static control system comprising procedural rules and regulations and utilising training manuals which may be bulky despite being limited in scope. The major disadvantage of too much reliance on such a means of control is that organisations run the risk that deviations in planned performance will remain undetected for some considerable time and that the effectiveness of responses to unsatisfactory performance, once detected, may not be readily ascertained.

A dynamic control system overcomes the disadvantages of a static system by being much more sensitive to deviations in planned performance. This enables appropriate responses to be implemented sooner. Performance can be ascertained under a dynamic control system by measurement of the extent to which goals are achieved. Without defined goals and measurable objectives performance may remain an intangible or at best subjective concept.

It is suggested that in practice few client organisations have the resources necessary to implement a dynamic system of control fully, so a control system based on a degree of compromise between the two available methods is the norm. Compromise might be evident in the application of dynamic control to small works operations with a high priority; those which are income-sensitive to the organisation, for instance. The remainder of small-scale building work might be carried out with less rigorous control exerted by rules and guidelines contained in the ubiquitous training manual. Perhaps in some organisations the best feedback is regarded as being silence from the person requesting the order in the first place. This is often considered to equate to tacit satisfaction with the job since if the job has not been done or has not been carried out with the required speed or quality the originator will usually react accordingly.

The costs of a control system may be significant so a fundamental requirement is that these should not exceed the savings that are possible from the appropriate level of supervision for the job. Control systems should incorporate the most appropriate measures for control of the task in hand. This recognises the fact that the effective control of quality has intrinsically different requirements from control of time. Effective control requires that sufficient appropriate control data are available on which to base assessments of performance leading to decisions aimed at correcting problems. Controls should be implemented and maintained as frequent as necessary to ensure the required level of performance. Simplicity of control mechanisms is desirable. This will facilitate their operation and accuracy. The operation of any control mechanism must not be beyond the capacity of the management structure and systems in place.

So what is actually being controlled? In small works procurement and management, like so many branches of management in general, the fundamental issue in most organisations is often expenditure. Time and quality are also significant factors, and in some cases time will be the prime factor.

These are the usual primary goals but, commonly, there will also be secondary goals. An appropriate measure of control is required to ensure that the realisation of the important goals for a particular job or process is being optimised. Where multiple goals exist for an item of work, some method of prioritising these is required. This in itself constitutes a means of exerting control.

Management thrust

Small works are manifest in various guises and, similarly, elements of the total workload in a large organisation will be financially contained in, and constrained by, many different budgets. However, leaving aside the discrete budgets and different subgroups of small works, it is possible to identify the management thrust of the main categories of small works mentioned in this book and defined in terms of the small works spectrum. This is a useful concept because the management thrust is related to management control, and if a common thread can be established for the bulk of work carried out within each of the three main subgroups of small works, this should point the way for the design and operation of the small works control system. Furthermore, the management thrust constitutes a logical starting point for the setting of goals and objectives for each of the subgroups of small works.

At the small-scale end of the small works spectrum there is little scope to reduce the cost of individual jobs so the management thrust will be directed towards reducing the frequency of occurrence of work in this category – this being the ethos behind planned preventative maintenance, for example. The incidence of jobbing works can certainly be reduced by implementing the right balance of planned maintenance work within the total workload, but the danger of carrying out too much planned work should be recognised (see Fig 8.1).

Much work of a jobbing nature, especially alterations and improvements, is instigated by building occupiers. They will probably be departmental managers and will in effect be 'customers' of the maintenance managers. Insofar as they set their priorities in terms of service standards for carrying out the work, they will have a great influence on the selection of procurement route. Unfortunately for the optimisation of efficiency, they may not be aware of the financial consequences of their decisions; nor will they be aware of the opportunity cost of the resources switched from carrying out other work to undertake the work they requested. By introducing budgetary accountability to occupiers and by providing them with feedback information on the costs of their requests for work, a

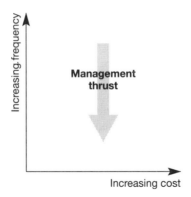

Figure 8.1 The management thrust towards reducing the frequency of occurrence at the small-scale end of the small works spectrum

climate should be fostered under which only absolutely necessary work should be carried out. Works prioritised as emergency, urgent and routine by these occupiers should normally reflect the increased costs associated with more urgent work on the feedback information provided to the customers. This should encourage them to categorise work at the lowest level of urgency and lowest cost.

Between jobbing works and minor projects lies ordered works, the boundaries of which are somewhat hard to define. As regards the management thrust, there will be a limited amount of scope for reducing the frequency of occurrence of work in this category and also for reducing the cost of individual jobs. There should, however, be sufficient flexibility in the setting of objectives for this category of work to enable either of these ends to be achieved where appropriate. Since the greatest proportion of work in the total workload is likely to fall into this category, this should reflect the management control and supervisory effort directed at this area of work (see Fig 8.2).

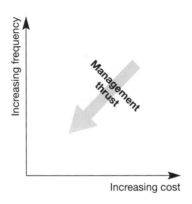

Figure 8.2 The management thrust towards reducing the frequency of occurrence and reducing cost in the ordered works region of the small works spectrum

Minor projects will usually be deemed to be of such importance that there will only be very limited scope for reducing the frequency of work in this category but there is likely to be much greater scope for minimising expenditure on each individual job (see Fig 8.3).

The concept of expressing control in terms of the different directions of management thrust for each of the three subgroups of small works is useful because the means of exerting control must take account of the management approach; that is, the basis upon which the goals, plans and actions are monitored and compared as the basis of a dynamic, feedback-based control system must reflect the differences in the management thrust. Ideally, management control procedures should be directed towards attempts to reduce the frequency of occurrence of jobbing works, reduce the cost of individual minor projects and reduce both the frequency of occurrence and cost of ordered works, where possible.

The importance of management thrust is frequently demonstrated in term contracts where routine small works are managed by term contract site superintendents, while larger works, when they occur, are taken outside the term contract organisation and managed separately, although they may use some existing staff, the same contractor and run parallel with the ongoing term contract.

Importance of supervision

It is the ethos of this book that the management of small works by large client organisations requires to have well-planned and well-defined goals and measurable objectives. It is only by ascertaining the extent to which these goals and objectives are realised can assessments of performance be made, decisions as to resource requirements made and effective forward planning undertaken.

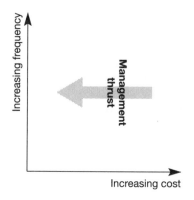

Figure 8.3 The management thrust towards reducing the cost at the project-scale end of the small works spectrum

Supervision and practical control of the workload should be mainly concerned with ensuring that the achievement of the specified goals and objectives, or critical success factors, is optimised.

The requirements for effective supervision and scope for control of the workload vary according to the various categories of small works identified in the small works spectrum. The high frequency of occurrence of small works jobs, particularly at the jobbing works end of the spectrum, means that close and thorough supervision of each and every job is impossible with the limited resources available to most maintenance managers. Effective prioritising is the only practical means of achieving optimal resource use.

Review and discussion

The costs of a control system may be significant, so a fundamental requirement is that these should not exceed the savings that are possible from the appropriate level of supervision for the job. Control systems should incorporate the most appropriate measures for control of the task in hand. This recognises the fact that the effective control of quality has intrinsically different requirements from the control of time.

Effective control requires that sufficient appropriate control data are available on which to base assessments of performance leading to decisions aimed at correcting problems. Controls should be implemented and maintained as frequently as necessary to ensure the required level of performance. Simplicity of control mechanisms is desirable, as this will facilitate their operation and accuracy. Furthermore, the operation of any control mechanism must not be beyond the capacity of the management structure and systems in place.

Since any control system is resource-intensive in itself, there must always be a cost/benefit analysis approach to control system design. What should be avoided is the situation where control system costs rise faster than the benefits of improved performance due to the existence of the control system. At the same time it is imperative that any management process that consumes the amount of resources that small works management does requires an appropriate degree of control to be exerted over it.

Fundamentally, whether the emphasis for control is on budgetary, quality, objective setting or any other aspect of management worthy of control, there is a necessity for mechanisms to measure, compare and effect appropriate responses wherever and whenever required. Given the high number of individual jobs existing within the total small works workload in the large organisation, maximum benefit can be achieved by employing computer systems to monitor performance.

In the quest to ensure that value for money is being obtained, benchmarking is in vogue as a means of providing a meaningful basis for comparative performance measures to take place. The key indicators of good small works management performance are identified for each relevant category, such as particular buildings, classes of work or areas of space occupied. Measures can be taken over time of the level of performance of each key indicator. Such measures may be cross-sectional, by comparing against levels of performance of similar buildings for instance, or may be longitudinal, by regularly measuring and comparing over time to detect changes in performance, but preferably will contain an element of both.

In all such assessments of performance, effectiveness of the resource use should be the watchword, not necessarily efficiency. Costs provide the obvious base for comparative measures but other meaningful key indicators of performance might include numbers of complaints of unsatisfactory performance, achievement of response times and management time spent on administration. The process of benchmarking compares and contrasts key performance indicators with those of peer groups so that optimum performance can be identified and the reasons for its achievement understood. The aim is then to seek continuous improvement in performance across both good performers and poor performers identified in the comparative process. Fundamentally, this is achieved by gaining as detailed an understanding as possible of the contributory factors that go to make up good performance, derived from analysis of basic data which possibly stems from completely unrelated industries or contexts, prior to internal application in the attempt to achieve greater performance.

Reference

1. Griffith A 1992 *Small building works management* Macmillan, Basingstoke

Chapter 9

Works policy: principal considerations

Workload and goals of the organisation

This book has dealt with most of the important aspects underpinning the design of a large client's small works procurement and management approach. This chapter seeks to summarise the salient points contained in the preceding text by way of basic considerations and outline guidance for practice.

In Chapter 1, reference was made to traditional ways of organising and managing the small works workload in the large organisational context. These were argued to be unsatisfactory in that they amount to little more than a means of procuring skills and labour from the marketplace. Management control in the conventional sense is rarely achieved to anything like a desirable extent and true measures of performance seldom exist. Without effective measurement of performance there can be no certainty that an organisation's small works strategy is the one that best supports the achievement of its goals. This is unsatisfactory in the face of ever more rapid organisational change in which small works must play its part by being managed in the most appropriate way to meet the challenges and changing needs of the parent organisation.

Chapter 2 sought to define small works and its component subgroups in definitive terms as a necessary precursor to the later exploration of the facets of its management. It was stressed that every client should fully understand the exact composition of the total workload, in terms of both distinct subgroups of small works and the generic types of work carried out.

- The client must be aware of the existence of discrete subgroups of work within the sphere of small works as each of these may require bespoke treatment during the procurement and management processes.
- Each item of work should be assessed in terms of its particular, and possibly unique, characteristics so that the most appropriate procurement and management procedures might be selected for it.
- A risk analysis exercise, cursory or thorough, should be conducted for each item of small works to aid the decision-making process as to the level of rigour to apply to its procurement and management and to assess the need for contingency plans.

Chapter 3 viewed small works in the context of the large organisation and emphasised the management concerns resulting from the large number of individual jobs. This was seen to produce an operations management situation at the small-scale end of the small works spectrum. Also highlighted was the potential for widespread inefficiencies in small works management to have a significant impact on the activities of the organisation.

Inefficiencies hardly matter on the individual job level but the aggregate effect of many hundreds of poorly managed jobs can be serious. Also emphasised was the importance of clearly defined goals and objectives in small works management. Both operation-oriented goals and project-specific goals should be formulated for small-scale and larger-scale small works respectively. All small works goals, and indeed all property goals, must stem from the organisation's business goals.

- The design of the small works procurement strategy should take account of the often routine nature of jobbing works and ordered works and the more unique nature of minor projects.
- It is essential that management information systems are in place that will facilitate the gathering of the required degree of data pertaining to the composition of the small works workload so that a detailed understanding of the volume of work may be gained.
- The workload should be broken down into the categories that are seen as important to the organisation while taking account of the distinction between the subgroups of small works.
- An accurate impression of exactly what happens under each small works procurement option should be gauged.
- The management time spent on administering each category of work under each procurement route should be recorded.
- There should be a means in place of identifying both good and poor performance.
- The indirect costs associated with small works management should be measured, as well as the direct costs.

Procurement and organisation

Chapter 4 highlighted some of the considerations in framing a small works procurement and management approach. Each of these considerations might be expressed as an organisational goal. The unique nature of each client organisation is important here since each will have its own needs and aspirations.

The design of the procurement approach should be unique to suit the requirements of the particular client and should take account of the following:

- Characteristics of the business.
- Functions of the building stock.

- Relationship of the building stock to the core business – manifest in how income-sensitive the individual buildings and their functions are.
- Any objectives stemming from the business plan.
- Assessment of which procurement routes are likely to help achieve the desired goals and objectives.
- Local market.
- Management structure in place.

Chapter 5 constitutes a brief treatment of the decision factors underpinning the decision to go for in-house or external procurement and management of the workload. Again, the importance of organisational goals in making this decision was emphasised.

- Where an element of directly employed labour is retained by a large organisation, the size of the establishment should be such that it does not exceed that which is required to cope with the reduced workload during troughs in the cyclic workload.
- The optimum size for the directly employed labour establishment should be arrived at by identifying the critical activities or operations upon which the estate's smooth running depends, whatever the prevailing economic conditions – i.e. those for which the consequences of inadequate small works and minor maintenance support would have an adverse effect on the organisation's continued success (generally comprising the activities and operations that are income-sensitive within the organisation). The level of direct labour provision should then be geared exclusively to the priority areas so identified.
- Where direct labour is employed and where health and safety legislation is not contravened, multi-skilled craftspeople or technicians will provide most flexibility. Regional differences will have an effect on the practicality of flexible labour utilisation since, in areas of strong trades union influence, rigid demarcation may be necessary.
- Effective management of direct labour requires the use of fairly sophisticated control and planning techniques which should include quantitative techniques for the measurement of output.
- A formal system for the appointment of consultants should be employed.

The decision to engage consultants and contractors will be made on the basis of:

- Needs of the organisation.
- Trade-off between the in-house and in-depth knowledge of the organisation versus the external broad-based experience gained by consultants and contractors.
- Need for increased flexibility.
- Need for a competitive element in service provision.
- Need for independent assessments as a basis for decision making.

Chapter 6 stressed that the choice of procurement route constitutes one of the most important influences on the ultimate success or failure of the individual job. Some routes will be more likely to lead to the achievement of organisational goals and objectives than others. Considerations for the design of the procurement approach may be:

- Are the client's requirements in terms of small works procurement understood in detail?
- To what extent is the composition of the total work order volume understood in detail?
- What are the specific requirements, in terms of a hierarchy of organisational objectives, for each component of the total work volume?
- Which statutory regulations must be met?
- How important are the considerations of accountability for decision making and security?
- To what extent is there a full understanding of what happens during the full process of procuring work under each of the available options?
- Which procurement options are most likely to lead to the achievement of the important goals and objectives?
- What geographical constraints apply?

Considerations for the selection of procurement option for a particular item of small works are:

- Is the job of a maintenance, alteration or new work nature?
- What is the approximate cost of the work?
- What are the attendant risks accompanying the job, stemming from both the job itself and also the consequences to the organisation of non-performance?
- What is the desired standard of quality?
- What time constraints apply/what is the urgency?
- What time is available for preparation of the documentation?
- Does minimising the direct cost of the work outweigh all other considerations?
- What is the complexity of the work?
- What is known of the scope of the work at the outset?
- What is the predictability of the work?
- What is the probability of changes being made to the scope of the work during its progress?
- Are there any geographic constraints?
- What are the characteristics of the local marketplace?
- Can the work be procured effectively with minimal client resource input?
- What financial considerations or constraints, such as year-end spend, apply?
- Is there a requirement to subcontract an element or elements of the work?

Small works should be procured on a jobbing basis where:

- The key characteristics of the work require a minimum degree of formality between client and contractor.

Small works should be procured on an ordered basis where:

- The key characteristics of the work requires a greater degree of formality between client and contractor than that needed for jobbing works (works may be one-off or be one of many ordered under a term contract).

Small works should be procured on a small/minor project basis where:

- The key characteristics of the work require a high degree of formality between the client and contractor.

The relationship of the key characteristics and categories of work were described in terms of the small works spectrum in Chapter 2.

Chapter 7 briefly presents the types of contractor available to carry out small works on behalf of the client. In small works, with its emphasis on the maintenance of long-term working relationships between clients and contractors, the choice of the right contractor is of paramount importance.

- Selection of the right contractor will make supervision and control less problematic than it otherwise might be.
- Contractors should be in no doubt as to the client's expectations in terms of performance required.
- Least problems in supervision will occur where the objectives of the contractor can be brought closer to those of the client – idealistic, perhaps, but a situation worth striving for nonetheless.
- Some problems in the client's supervision and control of contractors arise as a result of changes to the scope of the work after awarding the work to the contractor.
- Prior to commencing on site, the contractor should be provided with information that is as complete as practicable, commensurate with the job's characteristics. Too much information may prompt a contractor to inflate the price so there may be a fine balance to be achieved.

Management supervision and control

Chapter 8 is concerned with management supervision and control, in terms of ensuring that the goals and objectives are realised to their maximum extent. Deviations from planned performance must be identified earlier rather than later, and appropriate action taken.

Management supervision and control should focus on the key primary goals – time, cost and quality. Time management is concerned with three aspects:

- **Programming** – Planning the small works over a predetermined timeframe.
- **Progressing** – Monitoring the works and comparing with the plan.
- **Actioning** – Taking steps to redistribute resources to keep the works up to, or ahead of, the planned timeframe.

Cost management is concerned with three aspects:

- **Budgeting** – Planning cost expenditure on small works over the set timeframe.
- **Control** – Measuring the actual expenditure against the budget.
- **Forecasting** – Reviewing future costs based on actual costs and variations to budgeted costs.

Quality management is concerned with two aspects:

- **Standards** – Defining the quality of work expected.
- **Systems** – Putting in place procedures to monitor, assess and control work performance.

The difficulties of prioritising small works, discussed in Chapter 3, can make management supervision and control mechanisms difficult to formulate. This is exacerbated by the specific characteristics of small works, identified in Chapter 2: nature, scale, uncertainty, diversity, location, and risk, which means that small works can be difficult to plan for with certainty of success.

Notwithstanding the difficulties of small works management, a client should undertake the following tasks, some of which are non-recurrent and some routine:

- Present a clear statement of policy, goals and objectives for the management of the small works workload.
- Prepare a long-term programme and cost budget for the works.
- Classify the workload into categories, taking into account goals, objectives and the key characteristics of the works.
- Formulate the management strategy towards the monitoring, control, feedback and review of the categories of works.
- Prepare a medium-term programme and cost budget.
- Update the medium-term programme and cost budget based on actual performance.
- Prepare a short-term programme and cost budget.
- Monitor ongoing works by inspection.
- Inspect, approve, measure and agree the works completed.

- Update the short-term programme and cost budget, if essential to ongoing works or works which immediately follow, or adjust medium-term programme based on actual performance.
- Provide a review mechanism which considers the status of the total small works workload when required by management.

The mechanisms of management supervision and control should, ideally, meet the following requirements:

- Control should be based on clearly defined goals and measurable objectives.
- A means of sensing, comparing and effecting change is required for control to be effective.
- The design of the small works time/cost control systems must take account of the distinct subgroups of small works, their characteristics and the management thrust required.
- The level of control must be economically viable and the costs of the control system itself must not exceed the benefits of closer control.
- It is likely that there will be multiple control mechanisms in place, each dedicated to controlling a specific area, such as quality and budgets. These should be appropriate to their task and as simple as possible to achieve the desired ends.
- The identification and manipulation of trends is important for control systems. This requires sufficient basic data from which trends can be established and suitable responses planned.
- Control must be regularly, if not continuously, exercised since the need for appropriate responses to trends may otherwise be missed.
- The control system must be capable of being operated with the resources available.

Considerations for the design of the management approach:

- Have the policy, goals and objectives for the small works workload been determined?
- Has the workload been subdivided into categories of small works and have their specific requirements been considered?
- Have the key characteristics for each small works job been considered?
- Have long-term and medium-term programmes/cost budgets been determined?
- Are standard time/cost control techniques applicable or must particular mechanisms be developed and implemented?
- Have quality standards been determined and are mechanisms in place to monitor, control and report on work performance?
- Can the long-term and medium-term programmes/budgets be reconfigured into short-term control mechanisms?

- Are mechanisms in place to inspect, approve, measure, sign-off and agree payment for the works?
- Have measures been taken to ensure that each small works job is assessed and that the total small works workload can be evaluated periodically?

Review and discussion

The interrelated aspects of each of the foregoing chapters presents the basis for an overall small works procurement and management approach where a period of small works activity is planned. The paramount importance of goal-setting to the formulation of the small works management strategy should be immediately apparent. Goals and measurable objectives form the basis of performance assessments using a wide range of available methods.

Whereas control in many traditional small works management systems is exerted on a static basis by means of rules and detailed procedures set down in training manuals, the nature of control in the small works management approach is dynamic. Continuous feedback is the only way to ensure that adequate performance is being attained and that responses to change are timely and effective.

It is suggested that many current small works management systems simply do not lend themselves to the sort of management system advocated here. This is simply because the managers of small works become involved in the nitty-gritty activities of reacting to events, they need to authorise each job and inevitably lack sufficient time to manage the workload in the way that modern management techniques are applied to other processes in different industries. This by no means has to be so.

Why change things? Many managers of small works might hold the view that existing practices and procedures to procure and manage the workload have worked in the past and should continue to work well in the future, despite suggestions to the contrary. Such views would generally prefer to support the maintenance of the status quo rather than seek new and improved methods of managing small works. In answer to this, it is well to warn of the dangers of the unsophisticated image of building maintenance which is constantly perpetuated.

As stated in Chapter 1, the repair and maintenance sector of construction output is predicted to increase in importance in the future.[1] Consequently, proportionately more will be spent by client organisations on this sector than in recent times. It is essential that this expenditure is optimised. To achieve this, much more sophisticated management of clients' small works is required than clearly exists at present. Effective management is largely made possible by analysis of information gleaned from data gathering, and information technology will without doubt play a significant part in the future of small works management.

Many large clients' small works management procedures at present, however, simply do not lend themselves to data gathering and effective management control of the workload. *Ad hoc* systems and procedures are often the norm. In such a climate it is simply not possible to demonstrate the benefits of more sophisticated management techniques to the organisation's decision makers. The net result is that the importance of small works may be diminished in people's perceptions. Small works can become, therefore, somewhat marginalised and given perhaps less priority than other work, despite the fact that expenditure in the area is often great.

Small works and minor maintenance represent a very significant proportion of construction industry activity. It is therefore right, and indeed fundamentally important, to better appreciate this sector of work and recognise its significant contribution to any client's corporate effectiveness.

Reference

1. University of Reading 1988 *Building Britain 2001*

Select bibliography

Bowyer J 1976 *Small works contract documentation* Architectural Press, London

BMI 1990 *Measured term contracts* Building Maintenance Information, London

Campbell CW 1990 *The management and procurement of small works* MSc Degree thesis, Heriot-Watt University, Edinburgh (unpublished)

Centre for Strategic Studies in Construction 1988 *Building Britain 2001* University of Reading, Berkshire

Clamp H 1988 *The shorter forms of building contract* Blackwell, Oxford

Griffith A 1990 *Quality assurance in building* Macmillan, Basingstoke

Griffith A 1992 *Small building works management* Macmillan, Basingstoke

Griffith A and Headley JD 1995 Developing an effective approach to the procurement and management of small works within large client organisations. *Construction management and economics* **13:** 279–89.

Griffith A and Sidwell AC 1995 *Constructability in building and engineering projects* Macmillan, Basingstoke

Harlow PA 1985 *Managing building maintenance* Chartered Institute of Building, Ascot

Harris F and McCaffer R 1989 *Modern construction management* Granada, London

Lee R 1987 *Building maintenance management* Granada, London

McNulty AP 1982 *Management of small construction projects* McGraw-Hill, London

Milne RD 1985 *Building estate maintenance administration* E & FN Spon, London

Seeley IH 1987 *Building maintenance* Macmillan, Basingstoke

Turner A 1990 *Building procurement* Macmillan, Basingstoke

Index